THE WANDERING ASTRONOMER

THE
WANDERING
ASTRONOMER

PATRICK MOORE

CRC Press
Taylor & Francis Group
Boca Raton London New York

CRC Press is an imprint of the
Taylor & Francis Group, an **informa** business

CRC Press
Taylor & Francis Group
6000 Broken Sound Parkway NW, Suite 300
Boca Raton, FL 33487-2742

First issued in paperback 2019

ISBN-13: 978-0-7503-0693-5 (hbk)
ISBN-13: 978-0-367-39903-0 (pbk)

British Library Cataloging-in-Publication Data
A catalogue record for this book is available from the British Library

Cover design vy Kevin Lowry
Typeset in 12pt Garamond

Visit the Taylor & Francis Web site at
http://www.taylorandfrancis.com

and the CRC Press Web site at
http://www.crcpress.com

Author's Preface

Most books have a set plan. This one has not; it has no plan at all, and like its two predecessors, *Armchair Astronomy* and *Fireside Astronomy*, it is simply a collection of totally unconnected essays. Some are more technical than others but what I have tried to do is to present material which you will probably not find in conventional textbooks. Certainly it is not a book to be read straight through in the usual way; but if you dip into it at random, I hope you will find something that will appeal to you. I have done my best!

Patrick Moore
Selsey, June 1999

Contents

1

The Atmosphere of the Moon

"The Moon is an airless world." You will find this statement in countless books and, essentially, of course it is true enough. There is a very simple way to show that the Moon has at best a very thin atmosphere. When a star is occulted, it shines steadily right up to the moment when it is covered by the advancing lunar limb; there is no pre-immersion flickering or fading, as there is before occultation by Venus (and to a much lesser extent, Mars). But can we be confident that there is no atmosphere at all?

Originally, it was of course assumed that the Moon must have an atmosphere dense enough to support life. This was the firm belief of observers such as Johann Schröter, the first really great lunar observer, who began his work in the late 1770s. William Herschel, discoverer of Uranus and arguably the most skilled of all observers, believed the habitability of the Moon to be "an absolute certainty" (for good measure, he also believed in a cool,

Sir William Herschel, discoverer of Uranus.

inhabited Sun), and in 1822 Franz von Paula Gruithuisen announced that he had identified a true city with "dark gigantic ramparts", though, alas, there is nothing in this particular area other than low, haphazard ridges. This was in 1822; in the following decade many Americans were taken in by the celebrated Lunar Hoax, when the New York paper *Sun* announced that fantastic life-forms had been detected there by Sir John Herschel, who was busily surveying the southern skies from the Cape. (One earnest group even wrote to inquire whether there were any immediate plans to convert the Moon-men to Christianity.)

Low-type vegetation was still considered a possibility, even though a remote one, until less than a century ago, but so far as I know the last serious astronomer to believe in anything more advanced was W.H. Pickering, who made very notable contributions to lunar and planetary astronomy. Pickering observed an occultation of Jupiter, in 1892, and recorded a dark band crossing the planet's disk, tilted with respect to the usual surface belts. This he attributed to the absorbing effect of a lunar atmosphere. He repeated the observation at several later occultations, and found that the dark band was seen only when Jupiter was cut by the Moon's bright limb. At the dark or night side of the Moon it was never seen and Pickering concluded that the lunar atmosphere responsible for it was frozen solid during the Moon's night. He worked out that the ground density of the lunar atmosphere was about 1/1800 of the density of our own air at sea level.

But Pickering did not stop there. Between 1919 and 1924 he carried out a series of lunar observations from the clear skies of Jamaica, and in particular concentrated on the crater Eratosthenes, which has high, terraced walls and a central peak. Pickering claimed that dark patches inside it, which are easy telescopic objects, moved around during the lunar day and, although he was sure that vegetation tracts did exist on the Moon, he held that the spreading patches were better explained by swarms of insects. Finally, in 1924, he published a paper in which he claimed that the patches were probably due to much more advanced life-forms, which would make them even more curious. "While this suggestion of a round of lunar life may seem a

The Great Meteor of 7 October 1862 appears in this old painting. It was seen across England and apparently was much brighter than the full moon.

little fanciful, and the evidence on which it is founded frail, yet it is based strictly on the analogy of the migration of the fur-bearing seals of the Pribiloff Islands ... The distance involved is about twenty miles, and is completed in twelve days. This involves an average speed of six feet a minute, which, as we have seen, implies small animals."

Pickering died in 1938, and the idea of lunar creatures died with him, but the question of a residual lunar atmosphere was still carefully considered. In 1949 Bernard Lyot and Audouin Dollfus, both great French observers, used the fine 24-inch refractor at the high-altitude Pic du Midi Observatory, in the Pyrenees, to search for lunar twilight effects. They found none and concluded that any atmosphere must have a density less than 1/10,000 of ours. Next, V. Fesenkov and Y. N. Lipski, in what was then the USSR, carried out a search for twilight effects on the dark hemisphere of the Moon. Then in 1949 Lipski announced the detection of an atmosphere with a ground density 1/10,000 of that of the Earth's air at sea level. Unfortunately, confirmation from elsewhere was not forthcoming.

Observations were sometimes made of localised glows and obscurations on the Moon. These were known as TLP, or Transient Lunar Phenomena (a name for which I believe I was initially responsible). Because most of the observations came from amateurs, though not all, there was a good deal of official scepticism, but in 1999 the eminent French astronomer Audouin Dollfus, at Paris, obtained undeniable proof of a disturbance inside the great walled plain Langrenus; the evidence was conclusive. Dollfus was using very powerful equipment, and his work was a complete vindication of what the dedicated lunar observers had been saying for years.

Then, could any lunar atmosphere be dense enough to cause shooting star effects? After all, meteoroids become luminous in the Earth's upper air where the density is very low indeed.

In 1952 I had some correspondence about this with Ernst Öpik, an Estonian astronomer who spent much of his career at Armagh Observatory in Northern Ireland, and whose researches extended into many fields. I quote Öpik's letter to me, dated 11 September of that year:

"Lunar meteors are quite probable. Considering the surface gravity of the Moon, which leads to a six times' slower decrease of atmospheric density with height, the length and duration of path of meteor trails on the Moon will be six times that on the Earth, if a thin atmosphere exists. However, meteors the size of fireballs will penetrate the lunar atmosphere and hit the ground. The average duration of a meteor trail on the Moon will be two to three seconds (as against half a second for the Earth), and each trail should end with a flash when the meteor strikes the ground (because all meteors which can be observed in the lunar atmosphere, from such a distance, must be large fireballs). The average length of trail would be 75 miles (121 km) or one minute of arc — 1/30 of the Moon's diameter — and the meteors would therefore be very slow, short objects."

This sounded plausible enough and careful studies were made. In America, members of the Association of Lunar and Planetary Observers, directed by W. H. Haas, recorded quite a number of effects which met Öpik's conditions excellently; the average observed path-length was indeed 75 miles (121 km). In retrospect, this is rather significant. Since we now know that the Moon has no atmosphere which could possibly heat meteoroids to luminosity, what exactly did the American observers see? One fears that, as with the Martian canals, a good deal of wishful thinking must have been involved.

In the early 1960s, new attempts were made to track down traces of atmosphere by using the occultation technique (with radio, as well as visible light). Again the results were negative, and any atmospheric density had to be reduced to below 1/10,000,000,000 that of the Earth's air. It was only with the Apollo spacecraft that the first really reliable evidence came to hand.

Instruments carried on the orbiting sections of Apollos 15 and 16 detected small quantities of radon and polonium gas seeping out from below the visible crust. This was easily explained: radon and polonium are produced by the radioactive decay of uranium, of which there is plenty in the Moon's rocks. Indications of a very tenuous atmosphere were found by the instruments set up on the surface by the last four Apollos; the best results came from LACE, the Lunar Atmospheric Compositi Experiment, taken to the Moon by Apollo 17. The main gases turned out to be helium and argon. The helium came from the solar wind that strikes the Moon all the time; argon came from below the crust and was at its greatest concentration a few hours before each lunar sunrise. At this time the detecting equipment was much too cold for argon to be activated, and so it was concluded that a gentle argon wind was blowing over the horizon from the region beyond the terminator, which was warming up as the Sun's rays reached it.

Over a decade later, Drew Potter and Tom Morgan, using new equipment with the 107-inch reflector at the McDonald Observatory in Texas, identified two more gases, sodium and potassium.

The sodium appears to surround the Moon rather in the manner of a cometary coma; the atmosphere is surprisingly extensive, and its constituent atoms are moving around at great speeds, so that the "temperature" is high (though, of course, there is virtually no "heat"). It also seems that the entire atmosphere must be recycled and replaced every few weeks.

At least we now know the true situation. The total weight of the lunar atmosphere is of the order of 30 tons (30 tonnes); if it could be compressed to a density equal to that of our own air at sea level, it would do no more than fill a cube with a side length of 210 feet (64 m).

We have learned a great deal since Gruithuisen believed that he had seen a lunar city, and Pickering pictured animals wandering around inside Eratosthenes. Certainly the Moon does retain a trace of atmosphere, but so little that if we call our satellite an airless world we are not very far wrong.

2

The Man Who Discovered a Planet

In near-modern times there are only four astronomers who have been responsible for discovering new planets in the solar system. William Herschel happened upon Uranus in 1781, when he was undertaking a "review of the heavens" and was plotting stars in Gemini with the aid of a home-made telescope. In 1846 Johann Galle and Heinrich D'Arrest, from Berlin, identified Neptune; they had been given the position by the French mathematician Urbain Le Verrier. (They had had to obtain the permission of the Observatory director, Johann Encke, to use the Berlin 9-inch Fraunhofer refractor for the search. Encke's response has gone down in history: "Let us oblige the gentleman from Paris!") And then, in 1930, the ninth planet Pluto was tracked down by my old friend Clyde Tombaugh.

I first heard about the discovery of Pluto when I was seven years old, and at preparatory school. We had science once a week(!) and the discovery was mentioned more or less *en passant*. Little did I dream that fifty years later Tombaugh would invite me to collaborate with him in writing his book.

Tombaugh was born on 4 February 1906 at Streator, Illinois, where his family ran a farm. He was the eldest of six children; his family could not afford to send him to college, and he cycled to school, seven miles (11 km) and back each day. By the age of eleven he had already become a competent farm worker, but it was already clear that his interests were wide-ranging. History and geography fascinated him, and in 1918 a first view through his uncle's telescope turned his attention to astronomy.

In 1922 the Tombaughs moved to another farm, this time in Kansas, and it was here that Clyde made his first telescope; it had an 8-inch mirror, a wooden tube and a rudimentary mounting — but it worked, and worked well. Others followed and led

Pluto, the ninth planet from the Sun, was discovered by Clyde Tombaugh in 1930. Pluto is indicated by the arrows and the bright star is Wasat (Delta Geminorum).

on to systematic observation, mainly of the planet Mars. Diffidently, he sent some of his Mars drawings to V.M. Slipher, Director of the Lowell Observatory at Flagstaff in Arizona, and Slipher was impressed. He saw that this young man had exceptional talent, and the drawings arrived at exactly the right moment because Slipher was looking for an assistant to carry out a search for a new planet. It seemed that Tombaugh might well be suitable.

Neptune had been tracked down, in 1846, because of its gravitational effects on the motions of Uranus. Yet there was still something "not quite right" about the movements of the outer planets, and Percival Lowell, founder of the observatory at Flagstaff, had come to the conclusion that there ought to be another planet awaiting discovery. He worked out a position for it and searched, but with no success, and after his death in 1916 nothing more was done for some years — at least at Flagstaff. Abortive searches were carried on elsewhere, but not consistently. In 1928 Slipher decided to try again and acquired a fine 13-inch refracting telescope especially for the purpose. And in 1929 Clyde Tombaugh arrived at the Observatory to begin work.

The method was photographic. As we know, stars are so remote that their individual or "proper" motions are very slight, but a nearby object such as a planet will shift obviously against the starry background after even a night or two. Tombaugh began a systematic hunt. If two photographs of the same region were taken over an interval of several nights and then compared by using a device known as a blink-microscope, a planet would betray itself by its motion.

It was laborious work, but Tombaugh was both patient and skilful, and success was not long in coming. Images taken in January 1930 showed a dot of light which moved in just the correct way, and the hunt was over. Tombaugh examined the plates on 18 February, and saw that the object had to be moving well beyond the orbit of Neptune. In his own words:

"I walked down the hall to V.M. Slipher's office. Trying to control myself, I stepped into his office as nonchalantly as possible. He looked up from his desk work. 'Dr. Slipher,' I said, 'I have found your Planet X.' I had never come to report a mistaken planet suspect. He rose right up from his chair with an expression in his face of both elation and reservation. I said, 'I'll show you the evidence.'

"I explained that I had measured the shift to be consistent on the three plates, and that all the images were in the correct positions. Slipher kept flicking the shutter back and forth, studying the images. Then I said, 'The shift in my opinion indicates that the object is well beyond the orbit of Neptune.'

"Then Slipher said, 'Don't tell anyone until we follow it for a few weeks. This could be very hot news.'

"The excitement was intense. Another era for the Lowell Observatory was suddenly ushered in. The announcement three weeks later would cause excitement all over the world."

It did! The announcement was made on March 13, Lowell's birthday, and exactly 149 years after Herschel had identified Uranus. Suddenly Clyde Tombaugh became world-famous, and at that time, remember, he had no scientific degree of any kind. He was at once given an honorary degree; not surprisingly the university authorities refused to accept him as a member of a class studying elementary astronomy.

Clearly the planet had to be named. Various suggestions were made. T.J.J. See, formerly on the Lowell Observatory staff, recommended "Minerva", and this might have been accepted but for the fact that See was, to put it mildly, unpopular with his colleagues. Eventually "Pluto", suggested by an eleven-year-old Oxford girl, Venetia Burney, was the favoured choice, and it was appropriate enough; Pluto was the God of the Underworld and the planet named in his honour must be a decidedly gloomy place.

In fact, it proved to be an enigma. Its orbit was unlike that of any other planet, and swings it inside that of Neptune; moreover it is small and now known to be smaller even than the Moon. This means that it is of very low mass, and could not possibly exert any measurable perturbations on giants such as Uranus and Neptune. Either Lowell's reasonably accurate prediction was sheer luck else the real Planet X remains to be discovered. But this in no way detracts from Clyde Tombaugh's performance. Pluto was at that time below the fourteenth magnitude, and could so easily have been overlooked.

Tombaugh went on with his programme in the hope of finding yet another planet, and in fact he continued until 1943, by which time America had entered the war and all scientific programmes were disrupted. All in all he examined a total of 337 pairs of plates with the 13-inch refractor and several more taken with other telescopes. As he wrote later:

"Over 90,000 square degrees of plate surface were critically examined over every square millimetre, over a total plate area of 75.4 square metres, or 810 square feet. The total sky area is equal to 41,253 degrees; a total of 30,000 square degrees of the sky was blinked. From hundreds of small samples of star counts, the estimated number of stars in the examined areas totalled 44,675,000, or a total of 90,000,000 star images, counting those on each pair of plates. Every one of the 90 million images was seen individually by me. It required a total of 7000 hours of work at the Blink-Comparator.

"In this extensive search the following discoveries were made in order of importance: one trans-Neptunian planet, one globular star cluster, one cloud of galaxies, several lesser clusters of

The launch of a V-2 from Peenemünde. The V-2, a long-range liquid fuel rocket, was developed in Germany during World War II by Wernher von Braun.

galaxies, five 'open' galactic star clusters, one comet, and about 775 asteroids. On the plates, I marked 3969 asteroid images, 1807 variable stars, and counted 29,548 galaxies. It was an impressive record."

He left Flagstaff in 1944 and went to the White Sands proving ground in New Mexico, where the US Army was developing a launching base to test the V-2 rockets captured from the Germans. Tombaugh made valuable contributions, and developed new methods of optical tracking which are still in use today.

He found time to return to planetary observation, and made fresh studies of Mars; he predicted that the planet would have a surface pitted with craters, a prediction confirmed in 1965 when the first close-range images were obtained from Mariner 4. It was also from White Sands that Tombaugh undertook a systematic search for minor satellites of the Earth. He found none, and by now we may be confident that the Moon is our only natural satellite.

In 1958 he moved again, this time to become Professor of Astronomy at the New Mexico State University. It was there that he spent the rest of his career, remaining as Professor Emeritus even after his official retirement. He instituted a major programme of planetary research, and remained as its main organiser for many years. He was very active as a teacher, as well as an observer, and until the very last part of his life he was still to be seen at the University several days a week, working in his chaotically untidy study and giving help and advice to all those who asked for it. At his home in Las Cruces he set up a telescope, which looked unconventional but which was in fact a superb precision instrument. Typically, when I last saw the telescope — in 1995 — the cover of the secondary mirror was still a tin bearing the inscription: "Coffee".

I first came across Clyde Tombaugh in the 1960s, when I was working on one of the programmes in my BBC television series *The Sky at Night*. This particular episode was to deal with Pluto, so I contacted Clyde. He replied at once and we struck a friendship which, I am glad to say, proved to be permanent. He joined me in the programme on several occasions, and then paid me the great compliment of inviting me to co-author the "Pluto book". It was published in 1980*, fifty years after the great discovery. A special Pluto meeting was held at the New Mexico State University in Las Cruces and, of course, Clyde was the guest of honour. An asteroid, No. 1664, was named after him, and this was announced at the final banquet. His response was typical. "At least," he said, "I now have a piece of real estate that nobody can touch!"

He had married in 1934, and the union was ideally happy; the Tombaughs had a son and a daughter. Latterly the Tombaugh household also included a beloved black cat whose name was — of course! — Pluto.

Clyde Tombaugh died on 17 January 1997. He was a great man and a great friend, as well as a great astronomer. He is sadly missed, but will never be forgotten.

---

* *Out of the Darkness: The Planet Pluto*, by Clyde Tombaugh and Patrick Moore, published in the US by Stackpole Books and in Britain by Lutterworth Press. Unfortunately, now out of print.

3

Moon in Shadow

Nobody will pretend that an eclipse of the Moon is nearly as interesting or spectacular as an eclipse of the Sun. A total solar eclipse is unquestionably the grandest sight in all Nature; a lunar eclipse is a leisurely and gentle affair by comparison. All the same, it is always worth watching, and records of eclipses of the Moon go back for a long way. There are rather vague records of the eclipses of 3450 BC and 1361 BC and, of course, there are various Chinese records, mainly involving dragons.

Recently I came across an old book called *The Social Life of the Chinese*, written in 1867 by an author who rejoiced in the name of Rev. Justus Doolittle. I have no reason to doubt that he had done his homework well, and here is what he has to say about eclipse ceremonies:

"The high mandarins procure the aid of priests of the Taoist sect at their yamuns. These place an incense censer and two large candlesticks for holding red candles or tapers on a table in the principal reception room of the mandarin, or in the open space in front of it under the open heavens.

"At the commencement of the eclipse the tapers are lighted, and soon after the mandarin enters, dressed in his official robes. Taking some sticks of lighted incense in both hands, he makes his obeisance before or facing the table, raising and depressing the incense two or three times, according to the established fashion, before it is placed in the censer. Or sometimes the incense is lighted and put in the censer by one of the priests employed. The officer proceeds to perform the high ceremony of kneeling down three times, and knocking his head on the ground nine times. After this he rises from his knees. Large gongs and drums nearby are now beaten as loudly as possible. The priests begin to march slowly around the tables, reciting formulae, etc., which marching they keep up, with more or less intermissions, until the eclipse has passed off.

"A uniform result always follows these official efforts to save the Sun and the Moon. They are invariably successful. There is not a single instance recorded in the annals of the empire when the measures prescribed in instructions from the Emperor's astronomers at Peking, and correctly carried out in the provinces by the mandarins, have not resulted in a complete rescue of the object eclipsed."

The Greeks were rather better informed, and Anaxagoras of Clazomenae, who was born about 300 BC, was well aware that eclipses take place when the Moon passes into the Earth's shadow. *In The Clouds*, written around 410 BC, the dramatist Aristophanes refers to an eclipse seen from Athens on 9 October 425 BC. But there was one eclipse which had tragic consequences for the Athenians. In 413 BC the Peloponnesian War between Athens and Sparta was raging and the Athenian expedition to Sicily was in desperate trouble. Their commander, Nicias, should have pulled out his army by using the ships waiting in the harbour, but an impending lunar eclipse caused the astrologers to give him bad advice: Stay where you are for "thrice nine days". He did, and when he finally tried to get away, it was too late. Gylippus, the Spartan commander, had blocked his escape, his fleet was annihilated, and the whole expedition wiped out. That led to the subsequent defeat of Athens so that, but for the eclipse, the whole history of the Mediterranean — and possibly of the world — might have been different. Nicias was to blame; he ought to have known better. There was another military connection in September, 218 BC, when a lunar eclipse so alarmed the Gaulish mercenaries in the service of Attalus I of Pergamos that they refused to continue advancing into battle. Apparently this was not disastrous, and all in all Attalus seems to have been a fairly successful commander.

The Mesopotamians paid attention to lunar eclipses, and in a book by R. Campbell Thompson (*The Reports of the Magicians and Astrologers of Nineveh and Babylon*) we read: "When the Moon is eclipsed in Siwan [one of the lunar months] there will be flood and the product of the waters of the land will be abundant.

A lunar eclipse photographed by Cdr. H.R. Hatfield. Lunar eclipses only occur when the Moon passes into the shadow cast by the Earth. However, at most full moons, the Moon passes either above or below the shadow.

When at Siwan an eclipse of the morning watch occurs, Samas [the Sun god] will be hostile. When an eclipse happens in Siwan out of its time, an all-powerful king will die, and Ramman [the god of weather and storms] will inundate; a flood will come and Ramman will diminish the crops of the land, he that goes before the army will be slain."

A legend from Mongolia concerns a miscreant named Arakho, who was due to be punished for his misdeeds, but could not be found. When questioned, the Sun refused to give a straight answer, but the Moon betrayed him, so that he was found and soundly beaten. In revenge Arakho now chases the Sun and Moon over the sky. If he catches them, the result is an eclipse.

The Mexicans had their own way of dealing with the situation; they shot arrows at the eclipsed Moon. But come now to 1504, and Christopher Columbus. He was anchored off Jamaica and the local inhabitants were most unhelpful; they point-blank refused to supply him and his crew with provisions.

Unlike Nicias, Columbus knew a great deal about eclipses, and one was due. Accordingly he told the natives that if they persisted, he would make the Moon "change her colour and lose her light". The eclipse began on schedule, and the effects were immediate; Columbus was sent all the food he wanted, and there was no further trouble.

21

Then there was Captain Beeckman, commander of a ship anchored off an island near Borneo in November 1714. I quote his diary:

"We sat very merry till about eight at night, when, preparing to go to bed, we heard all of a sudden a most terrible outcry, mixed with squealing, hallooing, whooping, firing of guns, ringing and clattering of gongs or brass pans, that we were greatly startled, imagining nothing less but that the city was surprised by the rebels. I ran immediately to the door, where I found my old fat landlord roaring and whooping like a man gone raving mad. This increased my astonishment, and the noise was so great that I could neither be heard, nor get an answer to know what the matter was. At last I cried as loud as possibly I could to the old man to know the reason of this sad confusion and outcry, who in a great fright pointed up to the heavens and said 'Look there; see, the Devil is eating up the Moon!'"

Captain Beeckman knew more about eclipses than the famous author H. Rider Haggard. In his classic novel *King Solomon's Mines* he described a full Moon, an eclipse of the Sun, and another full Moon on successive days. When the error was pointed out, he altered the second edition, turning the solar eclipse into a lunar one.

Not all eclipses are equally dark. The French astronomer Danjon introduced an "eclipse scale" from 0 (very dark) to 4 (very bright, orange or coppery-red with a bright bluish shadow rim). He tried to link this with solar activity, eclipses being "dark" for two years after solar maximum, but it seems that the main factor is the state of the Earth's upper air, through which all light reaching us from the eclipsed Moon has to pass. There have been cases when the totally eclipsed Moon has disappeared so completely that it could not be found even with a telescope, as happened in 1761, according to the famous Swedish astronomer Per Wargentin (after whom the celebrated lunar plateau is named). Very often the colours seen during an eclipse are beautiful, and we can enjoy them, even if we have no chance to use an eclipse in the way that Columbus did.

4

The Zodiacal Intruder

The Earth moves round the Sun in a period of one year; therefore the Sun appears to go right round the sky in a period of one year. It crosses twelve constellations, those of the Zodiac, and

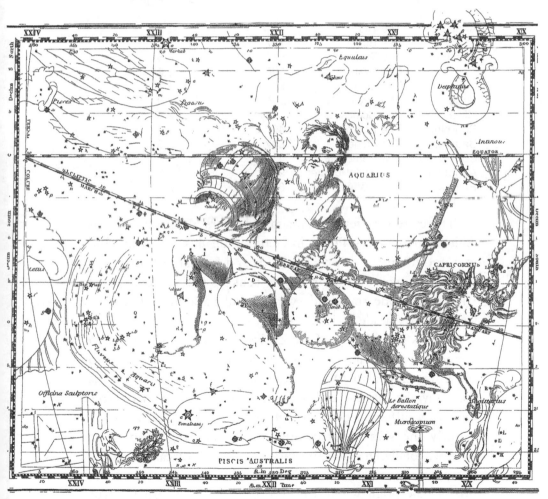

A star map drawn in the early 1800s shows the constellation of Aquarius, the Water-Bearer. Aquarius is one of the twelve Zodiacal constellations, all of which are named for animals and mythological figures. The Latin names given to the constellations by Ptolemy of Alexandria are still used today

this is where we can also find the Moon and the planets (apart from Pluto, whose eccentric path can take it well away from the Zodiacal band).

The twelve Zodiacal constellations are very unequal in size and in importance. Here they are:

Constellation		Area in square degrees	Number of stars above magnitude 5, per 100 square degrees
Aries	The Ram	441	2.5
Taurus	The Bull	797	5.5
Gemini	The Twins	514	4.5
Cancer	The Crab	506	1.2
Leo	The Lion	947	2.8
Virgo	The Virgin	1294	2.0
Libra	The Scales	538	2.4
Scorpius	The Scorpion	497	7.6
Sagittarius	The Archer	867	3.8
Capricornus	The Sea-Goat	414	3.9
Aquarius	The Water-bearer	980	3.2
Pisces	The Fishes	889	2.7

There are very few first-magnitude stars in the Zodiacal list; Aldebaran in Taurus, Pollux in Gemini, Regulus in Leo and Spica in Virgo, though Castor in Gemini and Shaula in Scorpius are only just excluded. For some reason the official list of first-magnitude stars ends with Regulus, at magnitude 1.35; Castor's magnitude is 1.58.

Scorpius is the richest of the Zodiacal constellations and Cancer the poorest. Of course, the constellation patterns are entirely arbitrary; we happen to follow the basic Greek system. If we had used, say, the Chinese or Egyptian pattern, our sky-maps would look very different, even though the stars would of course be exactly the same. Astrologers pay great attention to the Zodiac, totally ignoring the fact that the Zodiacal "signs" are not out of step with the actual constellations. The vernal equinox — the point where the Sun crosses the celestial equator each March, travelling from south to north — used to be in Aries, but has now shifted into the adjacent constellation of Pisces because of the effect of what is termed precession.

An astronomical chart of the Northern Hemisphere from Atlas Coelestis, published in Germany in 1742. It is a hand-coloured engraving started by Johann Baptiste Homann (1664-1724) and completed by Johann Gabriel Doppelmayr (1671-1750) in 1730.

25

In 1995 an astrological crisis blew up. According to headlines in many newspapers, astronomers had found a new constellation of the Zodiac, and hence a new Zodiacal sign! How would this affect horoscopes? And what influence would it have upon the unexpected arrival of tall, dark, handsome strangers!

In fact, the story originated in an astronomical broadcast on BBC television (not by me, I hasten to add!). Not that it was in any way new. The large constellation of Ophiuchus, the Serpent-bearer, sprawls across both the equator and the ecliptic, and therefore intrudes into the Zodiac between Scorpius and Sagittarius: planets may pass through it, and so does the Sun. As was pointed out in the broadcast, anyone born between 30 November and 17 December should really be said to be in the sign of Ophiuchus.

Ptolemy of Alexandria, last of the great astronomers of Classical times, left us a list of 88 constellations, all of which are still in use today, even though their boundaries have been somewhat modified. Ophiuchus was a "Ptolemaic" constellation, and so of course were the twelve members of the Zodiac; moreover Ptolemy was well aware of the phenomenon of precession. So the intrusion of Ophiuchus had been known for many centuries. Incidentally, there is yet another constellation — Cetus, the Whale — which is so close to the Zodiacal band that there are times when planets can pass through it, though the Sun does not.

On the whole, astrologers tend to ignore Ophiuchus, and pretend that it does not exist, but if they really concentrated I have no doubt that they could accommodate it, just as they managed to cope with the three planets not known in pre-telescopic times, Uranus, Neptune and Pluto.

5

Fishing Hipparcos Down

One of the more successful probes of the late 20th century was Hipparcos. It was launched on 8 August 1989 by the European Space Agency; its mission was to make accurate measurements of the positions of 118,000 stars. It was amazingly precise; the accuracy of the final measurements was from 1 to 1.5 milli-arc seconds, which is roughly equivalent to setting up a golf ball in New York and measuring its apparent diameter with an instrument in London (naturally, disregarding the curvature of the Earth!). There were initial problems because a launch malfunction meant that Hipparcos was put into the wrong orbit — an ellipse rather than a circle — but by the beginning of 1996 the catalogue was completed. It was published in the following year in seventeen vast volumes, which will be standard references for centuries to come.

Hipparcos had done its work, and done it well. It then ran out of power and was left in orbit. As it never comes much below 130 miles (209 km) it is "safe", and if left to its own devices will orbit the world for a very long time to come. But could we send up some sort of net and fish it down?

Let us admit that there would be only limited scientific value in this. Of course it would be fascinating to see Hipparcos in a museum, but the retrieval would cost a great deal of money, and whether it would be regarded as worthwhile is debatable. Certainly not at the present time, when funds are being cut to the bone and even experiments of immense scientific value are being either pruned or abandoned. But in the future things may be different. We may learn the art of spending less on nuclear submarines and more on constructive science, and Hipparcos would be a splendid prize.

Not long ago I was talking about this to Dr Floor van Leeuwen, then at the Royal Greenwich Observatory but formerly chief

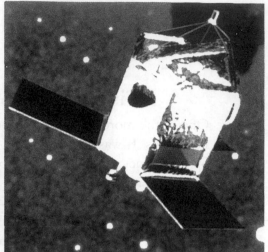

Hipparcos, one of the most successful astronomical satellites, was launched in 1989. Dedicated to the precise measurement of the positions, parallaxes and proper motions of the stars, it has collected high-quality scientific data.

project scientist of Hipparcos. No doubt he would like to see Hipparcos again, and I asked him for his views.

The orbit, as I have said, is elliptical. (It wasn't meant to be, but that is how it turned out.) This means that a rendezvous with it would be far from easy, because the relative speeds would be high, and there would be absolutely no chance of using the Shuttle. In fact, a manned retrieval seems to be out of the question. So if we are to capture Hipparcos, it must be by an automatic spacecraft.

What we really need is something in the nature of a rocket motor, more or less on its own. The trick then will be to launch it in the conventional way and put it into the same orbit as Hipparcos, which would be very difficult at the present stage of our technology but would presumably be fairly easy in a few decades' time. You then manoeuvre your "flying motor" until it edges up to Hipparcos. When they are sufficiently close together, you link them — possibly by an old-fashioned hook! — and then use the motor to bring the two craft gently down into a low, circular orbit. This will bring them within range of the Shuttle, or even of the international space station, which will by then be in orbit. After that, it would be a simple process to capture them and bring them home. After all, there have been two "captures" and subsequent releases of the Hubble Space Telescope.

I have a feeling that this may one day be done with Hipparcos, though because it is no longer transmitting we may very easily lose track of it (in which case, it would be very hard to relocate). Of course, there are other probes which we would very much like to retrieve, notably Giotto, which made its entry into Halley's Comet in 1986; but Giotto is in solar orbit, and the difficulties are much greater.

It is all very much up in the air, or rather, up in space, but at least Hipparcos will wait for us if we ever feel inclined to undertake a rescue mission. It has plenty of time to spare.

6

The Past and Future Moon

The Moon is moving away from us. Some people may be surprised to learn this, but I can assure you that it is true. Mind you, the rate of recession is not very great, and amounts to only 1½ inches (3.8 cm) per year, so there is no need to make haste to look at the Moon before it vanishes into the distance, but of its recession there is no doubt at all. At the same time, the average length of the Earth's rotation is increasing. The days are getting longer, though again, not by very much.

The reason is that the Moon raises tides in our oceans. (So does the Sun, but to a much lesser extent, so for the moment let us concentrate on the Moon). Tidal effects produce "drag", mainly felt in the shallow water separating Alaska from Asia, and the Earth's spin slows down, rather in the manner of a cycle-wheel spinning between two brake-shoes. But energy cannot be lost; it can only be transferred, and in this case it is transferred to the Moon, which is spiralling outward.

One way to prove this is by studying ancient eclipses. We know when these eclipses should have occurred, assuming that the Earth's rotation is constant; but there are discrepancies, and old records are quite clear. It is true that we now have atomic clocks

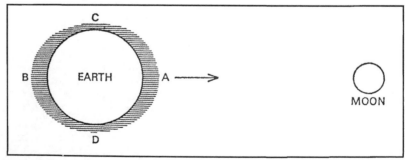

The Tides. If the Earth were surrounded by a uniform shell of water, there would be high tide at A and B and low tide at C and D. In fact the siituation is far more complicated than this!

which are better timekeepers than the Earth itself, and we know that the rotation period sometimes changes very slightly, but on average each day is 0.00000002 of a second longer than its predecessor. Let us see how this "secular acceleration of the Moon" shows up.

As each day is 0.00000002 of a second longer than the previous day, then a century (36,525 days) ago the length of the day was shorter by 0.00073 of a second. Taking the average between then and now, the length of the day was half of this value, or 0.00036 of a second shorter than at present. But since 36,525 days have passed by, the total error is 36,525 x 0.00036 = 13 seconds. Therefore the position of the Moon, when calculated back, will be in error; it will seem to have moved too far, i.e. too fast. It follows that eclipses of many centuries ago will not occur at the moments they would have done if there were no secular acceleration.

Going back many millions of years, the Moon must have been much closer to the Earth than it is now and our days will have been much shorter. Again we can draw evidence from the tides. In 1966 scientists in America, led by Charles P. Sonett and Aramais Zakharian, investigated layers of sediment deposited by ancient tides in the United States and southern Australia. They were mainly the Big Cottonwood Formation near Salt Lake City, deposited 900 million years ago, and the Elatina Formation near Adelaide, deposited 650 million years ago.

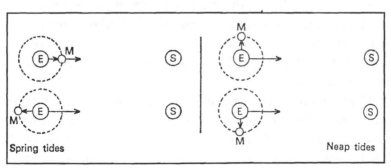

Spring tides   Neap tides

Spring and neap tides. Spring tides occur when the Sun and Moon are pulling in the same sense: neap tides when the pulls are in opposition to each other. In the diagram S stands for the Sun, E for the Earth and M the Moon.

31

Layers of tidally deposited sedimentary rock, called tidalites, are records of daily tides. Dark bands found in the layers mark the twice-monthly neap tides, when the Sun and the Moon are pulling against each other — that is to say, at half-moon — while lighter areas between the dark bands indicate the spring tides, when the Sun and Moon are pulling in the same sense. This, of course, happens when the Moon is either new or full. (The name has absolutely nothing to do with the season of spring.)

From their measures of the tidalites, Sonett and Zakharian found that 900 million years ago the day was only about 18 hours long. By the time of the Elatina Formation lay-down, this had increased appreciably, and has gone on lengthening ever since. The Earth's orbital period or "year" has not altered much, and so it can be calculated that in Proterozoic times, which ended around 600 million years ago, there were 481 days in a year.

We can also look into the future. If the present rate of recession is maintained, then in 15,000 million years' time the Moon will have moved out to a distance of 336,000 miles (540,000 km), as against the present value of 239,000 miles (384,000 km). The orbital period will then be 47 times as long as our present day, and this will also be the axial rotation period of the Earth, so that the Moon will remain fixed in the sky. Here let me quote a famous twentieth-century astronomer, Sir James Jeans, who wrote in 1929:

"The Moon is responsible for the greater part of the tides raised in the oceans of the Earth; these, exerting a pull on the solid earth underneath, slow down its speed of rotation, with the result that the day is continually lengthening, and will continue to do so until the Earth and Moon are rotating and revolving in complete unison."

"When, if ever, that time arrives, the Earth will continually turn the same face to the Moon, so that the inhabitants of one of the hemispheres of the Earth will never see the Moon at all, while the other side will be lighted by it every night. By this time the length of the day and the month will be identical, each being equal to about 47 of our present days. (Sir Harold) Jeffreys has calculated that this state of things is likely to be attained after about 50,000 million years."

"After this, tidal friction will no longer operate in the sense of driving the Moon further away from the Earth. The joint effect of solar and lunar tides will be to slow down the Earth's rotation still further, the Moon at the same time gradually lessening its distance from the Earth. When it has finally, after unthinkable ages, been dragged to within about 12,000 miles [19,000 km] of the Earth, the tides raised by the Earth in the solid body of the Moon will shatter the latter into fragments. These will form a system of tiny satellites revolving around the Earth in the same way as the particles of Saturn's rings revolve around Saturn, or as the asteroids revolve round the Sun."

A fascinating scenario, but it will not happen. Long before then, the Sun will have changed from a mild yellow star into a vast, bloated red giant, sending out at least 100 times as much radiation as it does now.

It is very unlikely that either the Earth or the Moon will survive, so that our life span is limited. Long before the Moon has time to spiral out to over 300,000 miles (483,000 km), both it and our world will have been destroyed.

Luckily, this will not happen yet awhile.

7

Visitor to the Ares Vallis

On 4 July 1997, America's Independence Day, a new spacecraft landed on Mars. Of course there had been many previous missions, but Pathfinder was unique. Instead of going into a preliminary orbit around Mars as all its predecessors had done, and attempting a "soft landing" by using parachutes and rocket braking, Pathfinder simply plunged headlong into the Martian atmosphere. Parachute braking was used, but the impact was anything but gentle. The probe was encased in air bags, which absorbed the main shock; it came down, bounced several times, and finally came to rest almost exactly where planned. Amazingly, it remained upright, and subsequently the ramps opened, allowing the tiny "rover", Sojourner, to crawl down on to the surface and begin a tour of exploration.

Sojourner was a tiny, wheeled craft; it could travel only slowly, and over a limited distance, but it worked well from the outset. It could negotiate minor obstacles and was skilfully guided by its controllers many millions of miles away. It was able to go up to the rocks and study them, sending back information about their nature and composition. And certainly there were rocks aplenty. The landing site took the form of an old floodplain at

The tiny "rover", Sojourner, was carried by Pathfinder, which moved around the Ares Vallis and sent back data about the rocks.

The Martian landscape. The rock in the right foreground is about 10 inches (25 cm) across. Most rocks appear to have vesicles, or small holes, in them. Such rocks on Earth can be produced by either volcanic processes or by hypervelocity impacts of meteorites.

the end of the Ares Vallis, which had undoubtedly been a raging torrent in the far-off days when Mars had liquid water. Rocks of all types had been swept down, some of which proved to contain a great deal of quartz. The rocks were given romantic names: Barnacle Bill, Yogi, Chimp, Wedge, Squash…. All had their own particular points of interest, and geologists were enthralled.

It was, of course, rather chilly on Mars; at noon the temperature rose above freezing point, but plummeted down to around -110°F (-79°C) at night, colder than anywhere on Earth. There was a gentle breeze, and an atmospheric pressure of 6.6 to 6.7 millibars. The daytime sky was pink, with tenuous icy clouds. There were even occasional dust devils, fortunately not strong enough to do any damage.

Obviously, Pathfinder and Sojourner could not continue to operate indefinitely. Their batteries were rechargeable, and could draw upon solar power, but it was not expected that Sojourner would be able to transmit for more than a week. In the event, data continued to come back until the end of September. Only then did the strange little machine, no larger than a television set, fall silent. And even before that happened the next probe, Global Surveyor, had been put into a closed path round Mars, preparing to begin a long, detailed programme of mapping.

Sojourner was not designed to search for life. That would be the task of a future mission, and final proof — or disproof — would come only with a sample-and-return probe. But it was a major step forward and the whole programme was justifiably regarded as a triumph. We know exactly where Sojourner is; no doubt it will be collected in the future and taken away to a Martian museum. Certainly it has an honoured place in history.

The idea of intelligent Martians had long since been discarded; Percival Lowell's brilliant canal-builders had been consigned to the realm of myth. And yet, how would we on Earth react if, suddenly, a strange object fell from the sky and began bouncing around, clearly on a tour of inspection? Let us allow our imagination to run riot. Picture the front page of the *Chryse Times* on the day of the landing. Perhaps it would have run as follows:

UFO over Chryse
Spacecraft or just a meteorite?

Statement from the Ares Vallis Observatory

Many reports have been received of a UFO over the central Chryse area. It is claimed that the pilot was seen to emerge, though the security authorities hastily cordoned off the area and no official information has been released.

A spokesman for the Tharsis UFO Society (TUFOS) states that the UFO almost certainly came from Bzork, the third planet in order of distance from the sun. The entry path indicates that the object is identical with a body detected telescopically some time ago and was found to be heading in our direction as though deliberately aimed. The approach velocity indicates that the UFO was launched from Bzork earlier in the year.

Dr Qrwalq, observing from the rim of Trouvelot Crater, told our reporter "The UFO moved very rapidly as it came down and made a loud swishing noise. As it dropped below a nearby hill there was a violent thud. Moments later the UFO reappeared, so that presumably the pilot had taken off in order to search for a more favourable landing site.

This happened several times. When the UFO landed I was able to train my binoculars on it through a fortunate gap in the hills. The upper section opened, and the pilot crawled out. I say 'crawled', but in a way the motion was more in the nature of a rolling gait, unlike any normal method of locomotion and more reminiscent of wheels. The pilot then moved over to a large rock and apparently began smelling it. Almost at once security officials moved in and erected a screen, so that I was unable to make any further observations.

However, a statement from Dr Zzyzz, Director of the Ares Vallis Observatory, discounts the UFO theory and attributes the phenomenon to a meteorite which fragmented during descent, thereby giving a false impression of movement after impact. The area, said Dr Zzyzz, has been cordoned off, because it is important to recover and analyse the meteorite before it can be vandalised by souvenir hunters. Bzork, he adds, is not a world which can possibly support advanced life. The temperature is far too high, the atmosphere is rich in the poisonous gas known as oxygen and there is almost continuous rainfall over the main landmasses. Life there, if it exists at all, must be very primitive.

Our reporter is still at the scene and is endeavouring to obtain further information. The object will be the subject of a special television programme this evening on Network Margaritifer.

One wonders what Lowell's canal-builders would have made of it.

8

Harvest Moons, Wolf Moons and Blue Moons

Look at the full Moon when it is low over the horizon and you may think that it seems huge — rather like a balloon floating in the sky. But when the full Moon is high up, it appears smaller. Or does it? In fact, the low-down full Moon is no larger than the high-up full Moon; the apparent enlargement is purely an illusion.

Many people find this hard to accept and the "Moon Illusion" has been noticed since very ancient times, but careful measurement will show that there is no difference at all. Seeing is not always believing.

It has also been claimed that "Harvest Moon", around the second part of September, looks larger than any other moon and rises at the same time for several consecutive nights. Wrong again, but this time there is a certain amount of truth in the suggestion.

The Moon is moving in its orbit from west to east, so that in the sky it seems to move from west to east against the stars, covering about 13 degrees per day. (Of course, this has nothing to do with the daily west-to-west motion, which is due solely to the axial rotation of the Earth. Remember, too, that the apparent diameter of the Moon is about half a degree.) The apparent path of the Moon in the sky is not very different from that of the Sun, since the angle between them is only five degrees, which is not much even though it is enough to prevent a solar eclipse from happening every month. When at full phase the Moon is opposite the Sun in the sky, and to observers in the Earth's northern hemisphere the Moon is then due south at midnight.

The fact that the Moon is moving more or less along the ecliptic at the rate of 13 degrees per day means that it rises later every night. This time-lapse is known as the retardation, and may sometimes amount to over an hour, so that if, say, the Moon

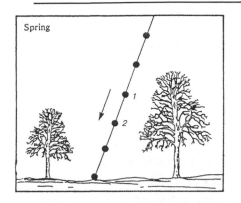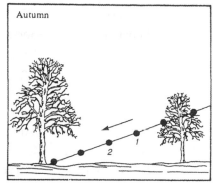

Harvest Moon. The Moon moves from 1 to 2 in the same time in each diagram. Obviously the retardation will be greatest in the spring (northern hemisphere; autumn in the southern hemisphere). This is the time of the harvest moon.

rises at 22.00 GMT on one evening it may not rise until 23.00 GMT on the next. But the amount of the retardation depends upon the angle which the ecliptic makes with the horizon, and this varies considerably. In the diagrams above, the Moon takes equal times to move from one black circle to the next. In the northern spring (around March) the angle is quite steep, and the retardation is at its maximum. In September, around the autumnal equinox (September 22) the angle is much shallower and it is easy to see that this cuts down the retardation. It is not correct to say, as many books do, that the full Moon then rises at the same time for several nights running, but certainly the retardation may be reduced to as little as a quarter of an hour. This means that it is at its best as a source of light, and farmers find it a help when gathering in their harvests, which is why the full Moon at this time of the year is called Harvest Moon. The following moon is often called Hunter's Moon.

Some time ago, following a BBC *Sky at Night* programme, I had a letter from a viewer asking me whether any other full moons had popular names. I did not know of any, and I am afraid I gave him the wrong answer. (Unfortunately, I did not keep a note of his address, so I cannot write to him to apologise.) When I started doing a little research, I soon found that other names were in use even though they were not widely known, and are restricted to folklore, often local folklore. Here are some of them:

January	Winter Moon, Wolf Moon
February	Snow Moon, Hunger Moon
March	Lenten Moon, Crow Moon
April	Egg Moon, Planter's Moon
May	Flower Moon, Milk Moon
June	Rose Moon, Strawberry Moon
July	Thunder Moon, Hay Moon
August	Grain Moon, Green Corn Moon
September	Harvest Moon, Fruit Moon
October	Hunter's Moon, Falling Leaves Moon
November	Frosty Moon, Freezing Moon
December	Christmas Moon, Long Night Moon

No mention here of a "Blue Moon", but there are two variants of this. First, there are times when the Moon really does look blue because of unusual conditions in the Earth's atmosphere. My best memory of this goes back to 26 September 1950, when I was living at East Grinstead in Sussex. I noted that "the Moon shone down from a slightly misty sky with a lovely shimmering blueness, like an electric glimmer, utterly different from anything I have seen before." It was also seen from other parts of the world and it was due to dust particles sent into the upper atmosphere by vast forest fires raging in Canada. The dust-pall in the American continent was really striking. At Ottawa, car headlights had to be switched on at midday and in New York a game of baseball was played under arc lights.

The synodic period of the Moon — that is to say, the interval between one new moon and the next, or one full moon and the next — is 29 days 12 hours 44 minutes 2.8 seconds, or 29.53 days. This means that in any month except February there may be two full moons, as in July 1998, when a full moon on the 1st was followed by another on the 31st. The second full moon in a calendar month is often called a blue moon, but I assure you this has absolutely nothing to do with colour.

9

The Cosmic Zebra

Is the zebra a black animal with white stripes or a white animal with black stripes? This, of course, is a question for zoologists, but there is an astronomical counterpart in the form of Saturn's eighth satellite, Iapetus.

Though always referred to as the eighth satellite, Iapetus was actually the second to be discovered, by the Italian astronomer G.D. Cassini in 1671, when he had been called to Paris to be

Iapetus, one of Saturn's largest satellites. An artistic impression by P.A. Helm.

Director of the new observatory there. (*En passant*, the design of the observatory itself was far from ideal; turrets and roofs blocked most of the sky, so that Cassini had to take his telescope into the garden!) At that time only one Saturnian satellite was known: Titan, found by Christiaan Huygens in 1655. Cassini went on to discover three more: Rhea in 1672, and Dione and Tethys in 1684. All these are visible under good conditions with a small telescope, and it has been claimed that people with keen eyesight can glimpse Titan with binoculars, though I admit that I have never been able to do so.

At this stage it may be helpful to list Saturn's major satellites, that is to say those known before the Space Age. Pan, Atlas, Prometheus, Pandora, Janus and Epimetheus have been added since, but all are small and so close to Saturn that they are very difficult to observe, and are quite beyond the range of telescopes of the sort generally used by amateurs. Data for the rest are:

Name	Mean distance from Saturn (miles/km)	Orbital Period d. h. m.	Diameter (miles/km)	Mean opposition magnitude
Mimas	115 (185)	0 22 37	248 (399)	12.9
Enceladus	148 (238)	1 8 53	310 (499)	11.8
Tethys	183 (295)	1 21 18	650 (1040)	10.3
Dione	235 (378)	2 17 41	696 (1120)	10.4
Rhea	328 (528)	4 12 25	950 (1529)	9.7
Titan	760 (1223)	15 22 41	3201 (5151)	8.3
Hyperion	920 (1481)	21 6 38	179 (288)	14.2
Iapetus	2200 (3541)	79 7 56	892 (1435)	var.
Phoebe	8050 (12,955)	550 10 50	136 (219)	16.5

Almost as soon as Cassini had identified Iapetus, he noticed something very strange. The magnitude varied in a more or less regular way. When the satellite was west of Saturn, it was easy to see, but when to the east of the planet it was much more difficult, and Cassini initially believed that it vanished altogether for at least a month during every revolution. More powerful telescopes showed that this was not so. At western elongation, the magnitude is at least 10, and probably brighter. William Herschel, in 1777, said that at its best it outshone Rhea, and this is also what I have found, so that my estimates give the maximum magnitude as about 9.5. But at eastern elongation, the magnitude dips to below 11, much fainter than Rhea, Tethys or Dione and not a great deal brighter than Enceladus (though Enceladus is much more elusive, because it is so much closer to Saturn).

The reason for this was soon found. Like nearly all major satellites in the solar system, Iapetus keeps the same face turned toward Saturn all the time, just as the Moon does with respect to the Earth. One hemisphere of Iapetus is as white as ice, and part of the other is as black as a blackboard; the albedo values are 0.5 for the bright side and only 0.05 for the dark side. When Iapetus is west of Saturn, it is the bright hemisphere which faces us, and this explains the variations in brightness perfectly. At one time it was suggested that the globe might be irregular in shape, but this did not seem plausible because Iapetus is relatively large, and most bodies of over around 500 miles (805 km) across are spherical. (Hyperion is shaped rather like a hamburger and has a chaotic rotation period, but of course Hyperion is very much smaller.)

This particular problem was cleared up in 1980, when the Voyager 1 spacecraft flew past Saturn and imaged Iapetus from a range of a mere 154,000,000 miles (248,000,000 km). In the following year Voyager 2 approached the satellite even more closely — 565,000 miles (909,000 km) — and sent back more detailed pictures. As expected, Iapetus turned out to be a cratered world, and the craters were given names from the Charlemagne period: Almeric, Othon, Roland and so on. The whole of the surface was cratered, so far as could be made out, but there was a very large dark region, now known (appropriately) as Cassini Regio. This occupied a large part of the less reflective hemisphere, which was the "leading" part of Iapetus as the satellite moved around Saturn. The line of demarcation was not abrupt, and there was a sort of transition zone from 130 to 190 miles (209 to 306 km) wide. It was also noted that some of the craters on the bright hemisphere had dark floors, with albedo similar to that of Cassini Regio.

No other satellite, either in Saturn's system or in any other, shows this peculiarity, so that Iapetus is unique. The first thing

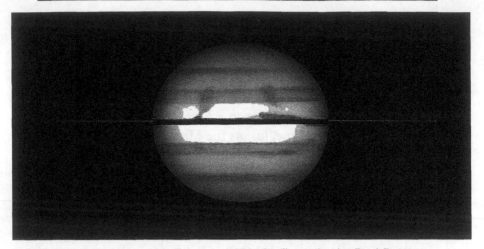

Saturn as seen on 24th February 1980. An illustration by Paul Doherty from a drawing by the author made with the 24-inch refractor at the Lowell Observatory.

to do was to decide whether the globe was bright with a dark deposit, or dark with an icy coating over the bright hemisphere. This was not difficult to decide, because the movements of Iapetus made it possible to work out the mass and, hence, the density. Overall, the mean density of the globe turned out to be no more than 1.16 times that of water, so that ice was clearly the major constituent. We had a white zebra with a black over-coat. Next, how did the dark deposit get there?

The dark region is on the "leading" edge, as we have noted, and this led to the suggestion that the black material might have been deposited there from space. Phoebe was one possible source. Here we have a much smaller body (discovered by W.H. Pickering in 1898) which is a very long way from Saturn and moves round the planet in a retrograde direction, in the manner of a car going the wrong way round a roundabout. It seems highly probable that Phoebe is a captured asteroid rather than a true satellite, and although it was not well imaged by either Voyager it was found to be darkish, with a surface coating rather like those of the so-called carbonaceous asteroids. Could it be that material was sputtered away from Phoebe and wafted inward, to fall on the leading hemisphere of Iapetus? This sounded con-vincing, but there are two fatal objections. One is that the

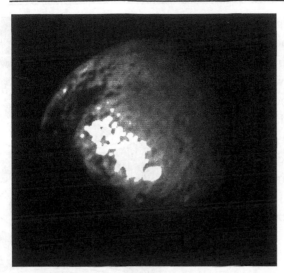

Iapetus, Saturn's outermost large satellite, photographed by Voyager 2 on 22nd August 1981 from a distance of 680,000 miles (1,100,000 km). Craters can be seen in both the light and dark regions.

colour and, therefore, the composition of Phoebe's outer layer is not the same as that of Cassini Regio on Iapetus. Secondly, craters on the bright area of Iapetus have dark floors; there are no "halo" craters in the dark region.

So far as we can tell, the dark regions are reddish, and the presence of basalts and silicates seems likely. Atmosphere, of course, is negligible at best — the escape velocity is only 0.42 of a mile (0.68 km) per second, so that although the bright surface is presumably coated with water ice there can certainly be no liquid water. All the indications are that the dark deposit is younger than the bright surface, and it is becoming more and more likely that the material has welled up from inside the globe, so that there must once have been considerable internal activity. We have no idea of the depth of the deposits; Carl Sagan suggested that there might be a very thick layer or organic substances.

For the moment this is about as far as we can go. We may know more in 2004, if the Cassini probe to Saturn is successful; the main aim of this mission is to drop the Huygens lander on to Titan and find out what the surface conditions there are like, but Iapetus will certainly not be neglected.

Will men ever reach Saturn's system ? If so, then the best views will certainly be obtained from Iapetus. All the inner satellites move virtually in the same plane as the rings, so that as seen from them the rings will always be edgewise on, but the orbit of Iapetus is tilted to the ring-plane by almost 15 degrees and the view will be glorious. Of course, this is looking far into the future, but one day it may well be that astronauts will stand among the Iapetan craters looking at the Ringed Planet from close range. Meanwhile, it is fair to say that Iapetus remains one of the most puzzling worlds in the entire solar system.

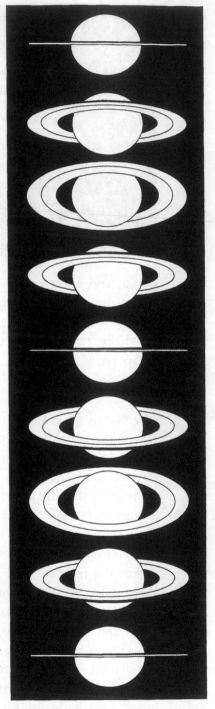

Saturn, the sixth planet from the Sun and the second largest in the solar system, is encircled by at least seventeen satellites and a complicated ring system. Illustrated here are the changing aspects of some of these rings. The inner satellites are best seen when the rings are edge-on.

10

Flying Saucers in Selsey

I do not believe in visiting spaceships flown by aliens from other planets. I have never done so, and I am sure that I never will. UFOs — Unidentified Flying Objects — have become all the rage, particularly in the United States, and there have been numerous reports of people who have been abducted by aliens and taken on joy rides into space. Of course, I do not for one moment suggest that we Earthlings are alone; there must be many civilisations in the Galaxy far more intelligent than we are, and certainly they could have mastered the secrets of interstellar travel and be quite capable of paying us a visit. Therefore, I am not denying the possibility of touring UFOs. What I do say is that there is not the slightest evidence that any of them have been here — yet.

However, there was one evening, years ago now, when my faith was temporarily shaken. I saw a whole flight of UFOs and, so far as I could tell, they were keeping me under close surveillance. It was quite unnerving, so let me begin at the beginning.

For much of my life I lived at East Grinstead, in Sussex, where I set up my modest observatory and concentrated upon mapping the Moon. I went briefly during the 1960s to Armagh, in Northern Ireland, to set up the new Planetarium there and remained for three years. After that, once the Planetarium was well and truly running, I came back to England and moved to Selsey, also in Sussex. My old thatched house is about five hundred yards from the sea, and by British standards the viewing conditions are fairly good since there is water on three sides of me and light pollution is less marked than in most other places. I acquired a 15-inch reflector, and set it up in an observatory, which looks remarkably like an oil-drum, but is extremely convenient and easy to use. A sleek, graceful dome is certainly much more attractive but mine, more on what may be called the wedding-cake pattern, is probably more "handy". The upper part moves round on rails, and the flaps are easy to open and shut. The telescope itself is on a fork mounting, conventionally driven; the optical system is pure Newtonian. Many modern amateurs have highly sophisticated, computer-controlled equipment, with all manner of electronic devices, including CCDs. I have not; I am old-fashioned and I still make direct observations at the eye-end of the telescope. I have been accused of being an astronomical dinosaur, and this is a charge which I am not prepared to deny.

The Moon has always been my main interest. I began lunar work when I was a boy and concentrated largely upon cartography, in particular what are termed the libration zones — the extreme edges of the Moon's disc, which are brought in and out of view because of the Moon's regular "wobble" from side to side. Obviously, features in the libration regions are highly foreshortened, and it is not easy to tell the difference between a ridge and a crater wall; before the Space Age our knowledge of these regions was decidedly sketchy. Maps were needed, and people such as myself did our best to provide them. Remember, before the flight of Russia's Luna 1, in 1959, we had never had any view of the "far side", the 41 per cent of the surface which is always turned away from the Earth.

The work was fascinating. I made a good many observations at the Lowell Observatory at Flagstaff in Arizona, using the great 24-inch refractor, and I also went to observatories such as Meudon, outside Paris, and the Republic Observatory in Johannesburg, but my own observatory was in constant use. The situation changed during the 1960s with the coming of the lunar probes, and our work became obsolete, as we had always known it would, but at the time it was useful. (I can modestly claim to have discovered the lunar Mare Orientale, which proved to be a vast feature of immense significance. Only a small part of it is ever visible from Earth, and even then only under ideal conditions. I was lucky enough to catch it at the right moment, but I had no idea of its true nature; I thought that it could be nothing more than a small "limb sea".)

The East Grinstead saucer. Experts claimed it to be from Mars. As the author was present (to the left) he knows that it was in fact an airborne soap dish!

Back to my UFOs. It was a clear morning, around 1 a.m., and I was ensconced in the dome. This was in the 1970s, and by then my main attention had switched to what are known as TLP or Transient Lunar Phenomena, minor outbreaks which seem to be due to gases escaping from below the crust and producing temporary obscurations. One area which is particularly subject to TLP activity is that of Aristarchus, a 23-mile (37-km) crater which is the brightest object on the whole of the Moon; it has regular walls, and a brilliant central peak. On this particular night it was well placed, and I put on a fairly high magnification of about × 400.

Suddenly I saw something remarkable. What appeared to be flying objects came into view. They moved fairly slowly and were fuzzy in outline; they did look remarkably like aircraft — but aircraft they certainly were not. I dashed outside, and saw nothing unusual. Back in the dome, there they were. I put on a much lower power, to give me a wider field of view, and they were quite striking. They shifted around, changing course abruptly and, for a while, they even seemed to keep in some sort of formation.

The dome of the 15-inch reflector in the author's garden.

I was frankly baffled. The silence was absolute, apart from the soft whirr of the telescope drive; the Moon remained clear, and so did the stars; the UFOs persisted. And then, after perhaps ten minutes, they started to disappear. They did not vanish instantaneously; they blanked out one by one, until there were no survivors.

My observing schedule for that night was hopelessly disrupted. I remained in the dome for a long time, until the sky began to brighten, but nothing more happened. As far as I was concerned, the mystery was complete.

Next day I did my best to think of a solution. There was no doubt at all that the UFOs had been there; they had been quite unmistakable and I had to admit that they did look exactly like aircraft (or spacecraft?) of some kind. It was tempting to believe that the Martians had arrived at last. I made some phone calls, and then, following a conversation with a botanist friend, I found the answer. At last I knew what the UFOs were — pollen.

Yes, pollen, drifting down in a light wind, catching the rays of the Moon as they came toward me. That explained the movements and the lack of sound; it also explained the fuzzy appearance. Since they were only a few yards away, my telescope naturally showed them not only magnified but also out of focus. Outside the dome, with the naked eye, I could not see them at all.

I felt relieved to have solved the problem, and also somewhat sheepish; I really ought to have put two and two together at the outset. Since then I have seen the same sort of effect on a couple of occasions but without harbouring any thoughts of marauding Martians.

Yet what would one do if confronted with a genuine UFO? On a broadcast discussion programme, I was once asked what I would say if a flying saucer landed in my garden and a little green man stepped out. I had the answer at once, and I knew what my opening gambit would be. "Good afternoon. Tea or coffee? Do join me — and then, do please come with me to the nearest television studio!"

11

Gene Shoemaker

On 19 July 1997, I was working quietly away at my study desk when the phone rang. I picked it up; it was a colleague from Australia. "We thought you ought to know straight away, Gene Shoemaker has been killed."

I think I dropped the telephone. It was an appalling shock, and I was utterly taken aback. Gene Shoemaker was an old and close friend; I had been a member of his lunar team in America, and physically he was nothing if not robust. He had been the victim of a tragic accident. He and his wife, Carolyn, had been in Australia studying impact craters. They were driving along the road near Alice Springs when they met another car head-on. Gene was killed instantly; Carolyn survived, though with broken ribs and serious bruising.

Eugene Merle Shoemaker was born in Los Angeles on 28 April 1928, but was brought up in Buffalo, in South Dakota. His father was a schoolmaster but later returned to Hollywood to work in the film industry. Gene was educated at the Californian Institute of Technology, gaining his bachelor's degree in 1947 and his master's degree in 1948. He joined the US Geological Survey and began searching for uranium deposits in Colorado and Utah, but already his thoughts were turning to the Moon and planets, and it was these which occupied his professional attention for the rest of his life. In 1951 he married Carolyn Spellman and this was one of the very happiest of marriages. Carolyn became Gene's scientific partner as well as his devoted wife.

In the 1950s he carried out an intensive study of Meteor Crater, Arizona, and it is said that eventually he knew it better than anyone else. Svante Arrhenius, an early twentieth-century Swedish physicist whose work was good enough to win him the Nobel Prize, once described it as "the most interesting place on earth", and certainly it is spectacular. Now it has become a well-known

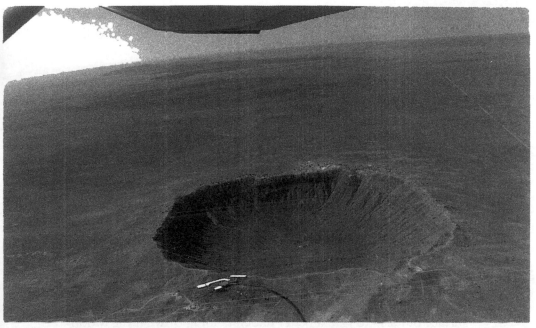

Meteor Crater, in Arizona, an aerial photograph taken by the author in 1997. The crater is about three-quarters of a mile (1.2 km) wide and scientists believe it was formed about 50,000 years ago by an iron meteor weighing more than 300,000 tons (tonnes).

tourist attraction, and is easy to locate, not far from the town of Winslow. Drive along Highway 99, turn off, and simply carry on until reaching the crater. Not that you will recognise it until you are almost up to it, because it is a sunken formation and although nearly 600 feet (183 m) deep there is not much in the way of a wall about the outer landscape. It is more than three-quarters of a mile (1.2 km) in diameter and the inner walls are quite steep, so that walking — or rather, scrambling — down to the floor is quite taxing. (I did it, years ago, but today the downward trail has been sealed off, to prevent accidents.)

It was first reported in 1871, and was tacitly assumed to be volcanic. Years later Daniel Moreau Barringer, a mining engineer, surveyed it and came to the conclusion that it had been produced by the fall of a meteorite, in which case the meteorite might still be buried underneath and could well be valuable. He brought in equipment and did his best to find the missile, but without success, and he gave up. We now know that he was correct in believing the crater to be of impact origin; the meteorite — or what is left of it — is almost certainly well below the south wall.

The Shoemakers confirmed that the crater was indeed an impact structure, and estimated its age as about 50,000 years. This was long before the area was inhabited, though the local Indians did tend to regard it with a certain amount of suspicion. Gene then turned his attention to other impact craters, of which there are quite a number. One of the best preserved is at Wolf Creek in Western Australia. It is not unlike Meteor Crater but is smaller, younger and much less easy to reach by land. At Henbury, also in Australia, there is a whole cluster of impact craters. You will find craters too in South Africa, Arabia, Argentina and elsewhere.

By 1960 it was becoming clear that the first manned expeditions to the Moon could not be long delayed, and it became essential not only to draw up really accurate maps of the lunar surface but also to gain a better understanding of what the Moon was really like. There had been considerable support for a rather strange theory that the lunar "seas" were covered with soft, deep dust, so that any spacecraft incautious enough to land there would sink without trace. Like most people, Gene was convinced that the Moon's craters were of impact origin, and he quickly became a leading expert in all branches of lunar study.

In 1961 he was placed in charge of a new scientific discipline, astro-geology, and based himself at Flagstaff in Arizona. He was very much to the fore with the successful unmanned missions of the 1960s — the Rangers, the Surveyors and the Orbiters — and it was his avowed ambition to make a personal trip to the Moon; after all, the presence of a geologist would be invaluable. In the event he did not pass the medical test, and he commented wryly, "Not going to the Moon and banging on it with my own hammer has been life's greatest disappointment." In fact, the only professional geologist who did go to the Moon was Harrison Schmitt, with Apollo 17 in 1972.

Gene left the Geological Survey in 1966 to continue with his own research, but he was still on hand to give what help was needed. I have good reason to know this, because at the time of

The Hoba West Meteorite near Grootfontein in Namibia, South West Africa, is the largest meteorite ever found. It weighs at least 60 tons (tonnes) and fell in prehistoric times; there is no crater.

the Apollo missions and the unmanned planetary missions I was very busy with television presentations, and in almost every case Gene joined for part or all of the time. He was untiring too in his investigations of impact craters and, together with Carolyn, he had also developed another field of research: comet-hunting and asteroid-hunting. To say that they were successful is an understatement. All in all they found 32 comets and no less than 1,125 asteroids. Much of their observational work was carried out at Palomar, in California, and it was from here, on 25 March 1993, that they made a particularly interesting discovery. Carolyn was examining one of the photographic plates taken earlier in the week when she noticed something very curious. "I don't know what this is," was her comment, "but it looks like a squashed comet." This turned out to be a reasonable description. The

55

comet was close to Jupiter, and was indeed in orbit round the Giant Planet; earlier it had passed so close in that it had been disrupted and had now been broken up into what has been described as "a string of beads".

For much of the research the Shoemakers had been joined by David Levy, a skilled amateur astronomer, and he had become a full partner in the research. He was with them on this occasion, and the "squashed comet" became officially Shoemaker-Levy 9, as it was the team's ninth discovery.

Calculations showed that SL9 was doomed. It was on a collision course with Jupiter and would hit the planet in July 1994. The excitement was intense, and telescopes all over the world were trained upon Jupiter because nobody was sure what would happen. If the comet's fragments were of appreciable size and rigidity, they would cause tremendous disturbances; if not, they could simply shower down almost unnoticed in the manner of shrapnel. Moreover, it was found that the actual impacts would occur on the side of Jupiter turned away from Earth, though the quick spin of the planet would bring the sites into view after only a few minutes.

I was observing from Herstmonceux Castle, in Sussex, with the 26-inch refractor. Herstmonceux had been the site of the Royal Greenwich Observatory but it was sold when the Observatory moved to Cambridge and the telescopes were left behind. We managed to rescue them and the great 26-inch was back in action in time for the SL9 impact. The first fragment of the comet impacted Jupiter around 20h 11m on 16 July and the last shortly after 08 hours GMT on 22 July. Each fragment produced a huge scar, and with the Herstmonceux telescope the effect was dramatic in the extreme, though the scars could be made out even with a 3-inch refractor. The results of the collision remained visible for months, and the whole encounter gave us a great deal of new information about Jupiter itself. Comet SL9, of course, was destroyed, but the Shoemakers still have many other comets to their credit and so, for that matter, has David Levy.

Another concern of Gene's involved what are termed NEAs,

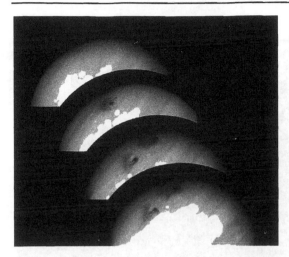

The comet Shoemaker-Levy 9 impacted with Jupiter between 16–23 July 1944. The comet was discovered on 25 March 1993 by Carolyn and Gene Shoemaker, assisted by David Levy.

or Near-Earth Asteroids. Though all the large asteroids keep strictly to that part of the Solar System between the paths of Mars and Jupiter, others do not, and there are small bodies which swing much closer in, so that they may pass very near the Earth. In fact there is always a danger of our being hit, and no doubt this has happened in the past (The sudden demise of the dinosaurs, about 65,000,000 years ago, has been put down to the impact of a meteorite, though I admit to being decidedly sceptical.) There are three "families" of NEAs; Amors (which cross the orbit of Mars but not that of the Earth), Apollos (which do cross the orbit of the Earth), and Atens (whose mean distance from the sun is less than ours). Up to modern times it had been thought that NEAs were uncommon. Gene disagreed, and so did Eleanor Helin, whose knowledge of asteroids is second to none. Systematic searches proved that NEAs are very plentiful, much more so than anyone could have expected, and by now several have been observed to pass between the Earth and the Moon.

If a NEA is seen to be on a collision course, can anything be done about it? A strike by a body of, say, half a mile (0.8 km) in diameter would cause an unbelievable amount of damage, and if we were hit by a ten-mile (16-km) asteroid there would probably be little left of civilisation. If we had enough warning, it is conceivable that we might be able to divert the asteroid by means

Herstmonceux Castle, once the headquarters of the Royal Greenwich Observatory, was sold when the Observatory moved to Cambridge.

of a nuclear warhead, but it is true that many NEAs are detected only when they have made their closest approach to us and are moving away again. The whole problem is now taken very seriously indeed, and the fact that we have been alerted is due mainly to the Shoemakers and to Eleanor Helin.

Gene was only 69 when he died. He will be remembered as one of the most cheerful and pleasant of men, and he is badly missed by his many friends all over the world. He and Carolyn will not be forgotten.

12

The Blackness of Mathilde

On 27 June 1997, the spacecraft NEAR (Near-Earth Asteroid Rendezvous probe) passed within 700 miles (1100 km) of the little asteroid 253 Mathilde, and sent back pictures which were, to put it mildly, surprising. Mathilde was simply not the sort of world that it had been expected to be.

Mathilde has been known for a long time. It was discovered on 12 November 1885 by the Austrian astronomer Johann Palisa, from Vienna Observatory. The orbit was computed by V.A. Lebeuf, then a staff member of the Paris Observatory. (He was born in 1859, and died in 1929.) It was also Lebeuf who suggested the name, which is said to honour Mathilde, wife of the French astronomer Moritz Loewy, who was at that time vice-director of the Paris Observatory. Mathilde's maiden name was Worms, so perhaps Asteroid 253 had a lucky escape!

Mathilde is a typical main belt asteroid in the midst of the swarm orbiting the sun between the paths of Mars and Jupiter. It has a period of 4.31 years, and the distance from the sun ranges between 181,000,000 miles (290,000,000 km) and 311,000,000 miles (500,000,000 km), so that the orbit is appreciably eccentric; the orbital inclination is 6.7 degrees, rather less than that of the planet Mercury. Mathilde is decidedly irregular in shape, with diameters of 31 × 31 × 43 miles (50 × 50 × 69 km). It shines as a dim star of around magnitude 11.9, so that it is beyond the range of most binoculars but is an easy object in a small telescope. From earth, of course, no surface detail can be seen, and even the Hubble Space Telescope can do no better, so that NEAR gave us our first reliable information about it. At the time of the encounter, Mathilde was 186,000,000 miles (300,000,000 km) from the sun.

The first surprise was the low density. The mass has been estimated at about a millionth that of the Moon and the density is

Two different views of Mathilde. The image on the left was obtained as the NEAR spacecraft approached Mathilde with its camera pointed near the direction of the Sun. Then as the spacecraft receded from Mathilde, it observed the asteroid almost fully lit by the Sun. Mathilde's irregular shape results from a long history of severe collisions with smaller asteroids.

only about 1.3 times that of water; most main-belt asteroids are much denser than that. For example, the density of Ceres, the senior member of the swarm, is 2.7 times that of water, and the brightest asteroid, Vesta, has a density of as much as 3.3, so that Mathilde is very different.

Next, it is very black. It was known to be a C-type or carbonaceous body, but the albedo or reflecting power turned out to be no more than 3 per cent, so that Mathilde is more than twice as dark as a lump of charcoal. Presumably it is made up of carbon-rich material which has not been altered by planet-building processes, which melt and mix up the original building block materials of the solar system. To quote one of the principal investigators, Dr Joseph Veverka:

"We knew that C-type asteroids are black, but we did not expect their surfaces to be as uniformly black and colourless as Mathilde's turned out to be. This global blackness is an important clue,

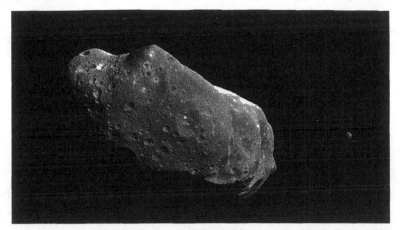

Asteroid Ida and its Moon — a picture transmitted to Earth from NASA's Galileo spacecraft. Ida, the large object, is about 35 miles (56 km) long. Ida's natural satellite is the small object on the right.

telling us that asteroids such as Mathilde are made of the same dark, black rock throughout, because none of the craters which are punched deep into the asteroid show evidence of any other kind of rock."

And craters abound there. The largest was found to be around 15 miles (24 km) across, and it was said at the time that "there are more huge craters than there is asteroid". Mathilde has had a tortured history, and no doubt its irregular shape is the result of collisions with other asteroids. All in all, it is not easy to see how it can have avoided being broken up altogether.

Finally, Mathilde is a very slow spinner. The rotation period is 418 hours, or about 17½ days, as against 9 hours for Ceres and only 5.3 hours for Vesta.

Mathilde is the third asteroid to have been imaged from close range; previously the Galileo probe, *en route* to Jupiter, had obtained pictures of Gaspra and Ida. Both these are irregular and cratered, but neither is as battered as Mathilde, and neither is as dark. Whether or not Mathilde is really exceptional remains to be seen, but certainly there is plenty to interest us in this curious, dusky little world.

13

Hitch-Hiker — and Others

Frankly, I cannot claim to have made many real contributions to astronomy, but I hope that my maps of the Moon were useful in pre-Apollo days. Anyway, the Russians in particular seemed to think so, but I played a minuscule rôle. I have, however, one claim to fame. I have introduced three astronomical terms which have now come into general use!

The first was pure chance. It was in the early days of the *Sky at Night* television programme, which began its monthly run in April 1957 and is still going. I was presenting a programme about the summer sky as seen from Britain, and this is, of course, dominated by three first-magnitude stars; Vega in Lyra, Altair in Aquila (the Eagle) and Deneb in Cygnus (the Swan). There is no connection between them, quite apart from the fact that they are in different constellations. Vega and Altair are normal Main Sequence stars, while Deneb is a celestial searchlight, shining some 70,000 times as powerfully as the Sun and so remote that we now see it as it used to be at the time of the Roman Occupation. Of the three Vega, almost overhead from England in summer evenings, looks much the brightest.

"These three are striking," I remember saying. "I think of them as making up the Summer Triangle." And somehow the expression caught on; everyone now refers to the Summer Triangle, but it must be added that that certainly does not apply to countries such as Australia and South Africa, where the Triangle is at its best during winter.

Next, the phase of Venus.

Because Venus is closer to the Sun than we are, on average 67,000,000 miles (108,000,000 km) from the Sun, as against our 93,000,000 miles (150,000,000 km), it shows phases like those of the Moon, from new to crescent, half, gibbous, full and back again to new. Obviously the phase should be easily

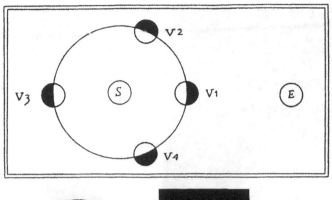

The phases of Venus. In the diagram, S represents the Sun, V1 to V4 Venus in different orbital positions: 1 (new), 2 (half-phase), 3 (full) and 4 (again half phase). Venus has its greatest apparent diameter when new and unobservable (except when it transits the Sun), and its least apparent diameter when full. When it is on the far side of the Sun it is again unobservable.

calculated; it should be 50 per cent — an exact half — when Venus is at right angles to the Sun with respect to the Earth. This is called dichotomy ("cut in half"). Surprisingly, the time of dichotomy never agrees with theory. During evening elongations when Venus is a narrowing disc, dichotomy is early; at morning elongations, when Venus is waxing, dichotomy is late. The amount of the discrepancy is not constant, but it can amount to two or three days. The exact moment of dichotomy is not easy to measure, because the terminator appears more or less straight for several consecutive days, but the effect is real enough.

It seems to have been first noted in the 1790s by the German observer Johann Hieronymus Schröter, who was an excellent

63

The William Herschel telescope on La Palma is Britain's largest. Photographed by the author in 1997.

observer and goes down as the first really great student of the Moon. In another programme I discussed it and referred to it as "the Schröter effect", and this too has been accepted by everyone.

There seems every reason to assume that the effect is due to Venus' atmosphere, which is very extensive; certainly there is not the slightest chance that the planet is out of position. Mercury does not seem to show the same phenomenon, simply because Mercury has practically no atmosphere at all.

Finally, come to April 1992, when I was joined in one of the programmes by Professor Michael Disney of Cardiff University, one of the world's leading astrophysicists. We were discussing the best ways of using telescope time, which is always at a premium. Michael Disney had been using the two main telescopes at the La Palma Observatory in the Canaries, the INT (Isaac Newton Telescope) and the WHT (William Herschel Telescope).

A large telescope can study, at most, only a handful of stars at one time, assuming that spectroscopic studies are to be made. Much of the other light received goes to waste. The Cardiff team had accordingly developed a new type of camera, which could be used at the same time as the main telescope. Once the telescope had been correctly aimed for a long photographic or spectroscopic exposure, the new mirror was made to glide into place, "stealing" all the light from the unused field of view, dividing it into four colours and collecting it by means of powerful video cameras. It has been said that during a typical observation, the camera increases the light-gathering ability of the telescope by a factor of about 3000. This makes it possible for British astronomers to make deep surveys of the remote galaxies at a cost of £12,000 per year; without this technique the cost would be more like £2,000,000.

The camera has been said to be the world's best for studying very faint, distant galaxies; it has been searching for what are nicknamed iceberg galaxies, which are so faint that their brightest parts are still below the darkest part of the night sky. It works well, and has done so from the outset.

During the programme, a thought came to me. "Well, this camera is merely tacked on to the main instrument, it gets in nobody's way, and it is a useful passenger. I suggest that it ought to be called 'the Hitch-Hiker'."

Michael Disney agreed. And the Hitch-Hiker it has remained ever since.

14

Ice on the Moon?

Everyone must have looked at the seas of the Moon. You can see them easily with the naked eye and there is no problem in making out the shapes of the Mare Imbrium (Sea of Showers), Oceanus Procellarum (Ocean of Storms), Mare Crisium (Sea of Crises) and the rest. But it has long been known that these romantic names are misleading. There is no water on the Moon; with the virtual absence of atmosphere the presence of liquid water is out of the question.

I remember a conversation held in 1967 between myself and Professor Harold Urey, one of the world's greatest geophysicists. We were attending the General Assembly of the International Astronomical Union, held that year in Prague, and one room had its floor covered with a vast lunar map. Standing on that chart, with one foot in the Bay of Rainbows and the other in the Sea of Gold, Urey told me that in his view the seas had once been water-filled. I was frankly unconvinced; certainly the seas were molten once but, surely, containing lava rather than H_2O. We agreed to differ.

Within a few years of that meeting the Apollo astronauts and the Russian unmanned probes had brought home samples of the Moon. Laboratory analysis showed a total lack of any hydrated materials; the Moon was bone-dry. That seemed to be conclusive, and water, in any form, was ruled out — until 1994.

In the first months of that year a new lunar spacecraft, Clementine, orbited the Moon and carried out a superb survey of the entire surface. In particular, it provided improved maps of the Polar Regions, which are never well seen from Earth because they are so foreshortened. Of course they had been studied from the various space missions, but Clementine's results were the best yet. And there was one real surprise. Clementine was said to have detected ice-lakes in some of the polar craters.

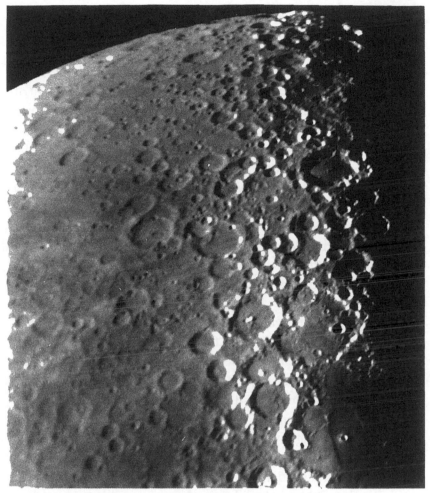

*The lunar craters near the lunar south pole (top of picture) have perma-
nently shadowed floors. The linear feature to the lower right is the Straight
Wall, so named because it is not straight and not a wall — it is a fault in
the surface!*

It so happens that near the poles, particularly the South Pole,
there are some large, deep craters whose floors are never free
from shadow and therefore remain bitterly cold. I would suggest
that these crater-floors must be the loneliest and most forbid-
ding places in the entire Solar System; they never see the Sun,
and on the airless Moon there is no twilight or diffusion of light.
So far as temperature is concerned, ice could exist there with no
trouble at all, but how could it be detected?

The answer is "by radar". Clementine carried highly sophisticated radar equipment, and one lunar scientist, Stewart Nozette, proposed to direct radar beams right into the craters. The pulses of energy bounced back for analysis might, he claimed, show whether or not ice existed. Three years earlier the same sort of experiment had been carried out with the planet Mercury, not from a probe, of course, but with the giant radio "dish" at Arecibo in Puerto Rico. This has been built in a natural hollow in the ground and is 1000 feet (305 m) in diameter. Arecibo scientists had found that permanently darkened polar craters on Mercury reflected radio pulses just as ice might be expected to do, and though many people were highly sceptical the investigation was certainly worth following up. If ice existed on Mercury, then why should it not exist on the Moon too?

Nozette's plan was put into action. There were a number of suitable areas, notably craters in the region of the vast South Pole-Aitken Basin, which has a diameter of over 1500 miles (2400 km). During April 1994, Clementine's path took it directly over the Basin, and this presented an excellent opportunity because ice reflects radar pulses very strongly when the angle between the transmitter and the receiver, as seen from the target, is close to 0 degrees. Pulses reflected from ice would also be expected to be slightly different from light sent back from non-icy areas, because of what are termed polarisation characteristics. To the gratification of Nozette's team, the results seemed to be positive. "Not only did the echo prove to be stronger than expected," wrote Nozette, "but the all-important polarisation ratio also showed a modest peak precisely when the shadowed areas were within the radar beam." It was claimed that the patches of ice, sandwiched in between areas of normal rocky surface, would add up to a slab the size of a football field and roughly 50 feet (15 m) high.

Of course, the announcement sparked off tremendous interest. It was suggested that the ice might be able to provide enough water to supply future colonies, for example. One NASA scientist was positively ecstatic. "With water you could have enclosed areas to grow plants, grow your own food, make your own fuel,

Lunar rover vehicles considerably extended the areas of exploration on the Moon. Apollo 17, the last lunar manned flight, was launched on 7 December 1972.

make your own air. You don't have to launch all that stuff from big rockets on the Earth." It all sounded most encouraging but there were various other points to be considered.

For example, there could hardly be much ice. It could survive only in the shadowed craters, under temperatures of around -380°F (-230°C). At best it could provide only enough water to fill a couple of good-sized swimming pools. Secondly, how could the ice have originated? Certainly it could not be truly "lunar", so that presumably it would have had to be dumped on the Moon by an impacting comet.

Most astronomers now believe that the lunar craters are impact structures rather than volcanic calderas. (I say "most" because there are a few dissentients, and I happen to be one, but that is another story.) Yet the impact of a comet would cause tremendous heat, albeit briefly, and any ice would be vaporised. The only way around this difficulty would be to assume that the impact produced a cloud of material, which subsequently condensed out in the only regions where the temperature was low enough. My instinctive reaction was that the material would be more likely to leave the Moon altogether; after all, the lunar escape velocity is a mere 1½ miles (2.4 km) per second.

Clementine left lunar orbit after a few weeks. (It was scheduled to go on to a rendezvous with an asteroid, Geographos, but faults developed in the spacecraft and this part of the programme had to be abandoned.) It was then recalled that radar studies of the polar areas had already been made with the Arecibo telescope, notably on 18 August 1992, when as seen from the Moon's south pole Arecibo had been just over 6 degrees above the horizon. The same effects had been recorded, but they had also been found in regions such as the Sinus Iridum, well away from the pole, where the daytime temperature is very high and ice could not possibly survive. The Arecibo team concluded "The coincidence of some of these features with the radar-facing slopes of craters and their presence in sunlit areas suggests that very rough surfaces rather than ice deposits are responsible for their unusual radar properties."

There the matter rests, at least for the moment. I remain sceptical; I do not believe that ice exists either on Mercury or in the polar craters of the Moon. There are no skating rinks in the lunar world.

15

Mira Stella

On 13 August 1596, a Dutch amateur astronomer named David Fabricius was looking at the night sky when he saw a reasonably bright star in the constellation of Cetus, the Whale. He paid little attention to it — why should he? — but a few weeks later it had disappeared. For some reason or other he did not look for it again, and the next sighting came in 1603, when Johann Bayer was compiling a new star-map and giving the stars their Greek letters which we still use today. There was Fabricius' star, and Bayer gave it the letter Omicron, so that it became Omicron Ceti. Once more it vanished. Finally, in 1638, another Dutchman, Phocylides Holwarda, established that it appears and fades regularly; it was the first known variable star. It reaches maximum

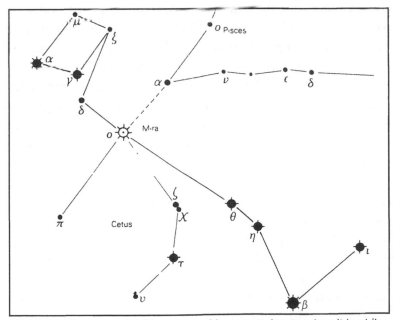

Mira is in a barren part of the sky and is easy to locate when it is at its brightest. Like all red giants, it has a diameter of more than 100 million miles.

71

every 332 days, and may then be of any magnitude between 2 and 4. When it drops to minimum, it becomes far too faint to be seen with the naked eye, or even ordinary binoculars, but it never falls below about magnitude 10, so that a moderately powerful telescope will keep it in view. On average, it is visible with the naked eye for several weeks every year. Hevelius, an early telescopic pioneer, nicknamed it Mira Stella, the "Wonderful Star", and we still call it Mira.

Mira is a pulsating star. It is a red giant, and it swells and shrinks, changing its output of energy as it does so; hence the fluctuations in light. It is now known to be the prototype of a whole class of similar stars, and there are several others that can reach naked-eye visibility, notably Chi Cygni in the Swan, R Hydrae in the Watersnake and R Leonis in the Lion. However, Mira is much the brightest of them, even though its distance from us a full 400 light-years.

Mira variables are characterised by the fact that neither the periods nor the amplitudes are constant. With Mira, the interval between successive maxima may be several days either longer or shorter than the mean value of 332 days. At some maxima the star may become really bright; in 1779 it is said to have reached the first magnitude, almost equal to Aldebaran, and I remember that in February 1987 I made it of magnitude 2.3, more or less equal to the Pole Star. However, at other maxima the greatest brightness may be as low as magnitude 4. One can never tell; no two cycles are exactly alike. Yet it is always easy enough to identify because it is so obviously red; its spectrum is of type M.

It is not alone in space. In 1918 A.H. Joy found certain peculiarities in the spectrum which indicated the existence of a faint but hot companion, and in 1923 the companion was discovered visually by R.G. Aitken, using the great 36-inch refractor at the Lick Observatory. The companion was not particularly dim, around magnitude 9.5, but was not easy to see because its apparent separation from the red star was only about 0.9 of a second of arc. It is even less easy now, because the separation has decreased, not because it has actually done so but because we are seeing the pair from a less favourable angle.

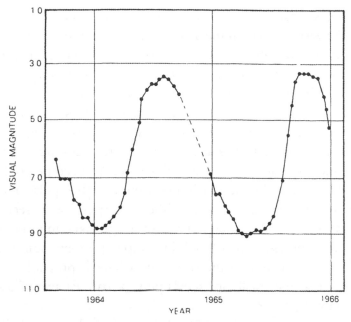

Mira is a long-period variable that is visible to the naked eye. The average period is 332 days.

The separation will be at its least, about 0.1 of a second of arc, in the year 2000, after which the pair will start to open out again. We are dealing with a binary system; the orbital period is around 400 years, and the average separation between the two stars is about 6500 million miles (10,460 million km), though the orbits are rather eccentric. The companion is now known to be a very dense, very small white dwarf, and it fluctuates irregularly in light. It is what is known as a flare star, and has been given a variable star designation — VZ Ceti.

In 1997 the Hubble Space Telescope was used to make a careful study of the Mira system and the results were fascinating. It is usually said that stars appear simply as points of light, but a few have apparent diameters large enough to be measured with modern instruments, and Mira is one of these. The star with the largest apparent diameter was known to be another red variable, R Doradûs, at 57 milliarc seconds; Mira proved to be about the same. But remember, it is 400 light-years away and the real diameter must be about 600,000,000 miles (965,000,000 km). This is large enough to swallow up the entire orbit of Mars round

73

our Sun, so that Mira is indeed a giant in every sense of the term.

It is a very old star, in the sense that it has run through a good part of its evolutionary cycle. It must once have been a star not too unlike the Sun, shining because of nuclear reactions going on inside it. Hydrogen was the main "fuel", and was being converted to helium, with release of energy and loss of mass. Unlike the Sun, Mira has run out of available hydrogen, so that different reactions have started. The core temperature has rocketed, and the outer layers have blown out to become cool and red; this is also why the star has become unstable, pulsating in a more or less regular manner. Neither is it spherical; it seems to be shaped more like a rugger ball. An image taken in ultra-violet light also shows a small, hook-like appendage extending from Mira in the direction of the companion. This is probably due to the fact that the small, hot star is pulling material away from its bloated red primary.

The companion is of a totally different type. In its evolution it is even more advanced than Mira; it has used up all its nuclear fuel, and its atoms have been broken and squashed together, so that the density is thousands of times that of water. It is still shining, but only dimly and there is considerable flare activity, so that the magnitude fluctuates between 9½ and 12. Eventually it will lose all its light and heat, and become a cold, dead globe — a black dwarf — this will not happen for thousands of millions of years yet, by which time Mira itself will also have collapsed into the white dwarf state. The difference in mass between the two components is not nearly so great as might be thought, despite the discrepancy in size. A small star is always denser than a large one, so that it is rather like balancing a lead pellet against a meringue.

It is always worth looking at Mira — with the naked eye when near maximum, or with a telescope when dim. It lies near the Whale's "head", as shown in the map; suitable comparison stars are Menkar or Alpha Ceti (magnitude 2.5), Gamma Ceti (3.5), Delta Ceti (4.1) and Alpha Piscium (3.8). Hevelius was fully justified in naming it "the Wonderful Star".

74

16

The Case of the Vanishing Planets

Our Sun is one of 100,000 million stars in the Galaxy and there
is nothing particularly distinguished about it. It is the centre of a
system of planets, so why should not other stars have planet
families of their own? There seems no reason why not. The trouble is that as yet we cannot see them; not even the Hubble Space
Telescope can so we have to rely upon indirect methods of detection. Ironically, the first three attempts ended in abject failure.

The individual or "proper" motions of stars are slight but easily measurable. For example Sirius has a yearly proper motion of
1.21 arc seconds, and as early as the eighteenth century Edmond
Halley realised that its position had shifted since ancient times.
Later it was found that instead of moving smoothly, Sirius was
"weaving" its way along, so that clearly it was being influenced
by a companion close beside it. In 1862 the Companion was
found; it proved to be a very small, very dense white dwarf star
with a diameter of a mere 25,000 miles (40,000 km) or so (comparable with a planet such as Neptune) but a mass equal to that
of the Sun. It turned up just where it had been expected. Years
earlier, the German astronomer F.W. Bessel had predicted where
it ought to be. Other white dwarf companions of bright stars
have since been found by this "astrometric" method.

But a white dwarf star is much more massive than any planet
could possibly be, and the astrometric method could be used
only to locate a heavyweight planet associated with a lightweight
star. During the 1940s slight "wobbles'" were suspected with
several dim stars, but were not confirmed, and the first positive
evidence came — or seemed to come — only with the work of
Peter van de Kamp at the Sproule Observatory, in Pennsylvania,
from 1937.

The star concerned was a faint red dwarf in the constellation
of Ophiuchus, to which attention was first drawn in 1916 by

Three images of Sirius and Sirius B. Sirius B (the Pup) is the dot below the over-exposed Sirius. Were it not so drowned by the brilliant light of the primary, the Pup would be a very easy telescopic subject.

E.E. Barnard; it is always referred to as Barnard's Star. Its position is R.A. 17h 55m.4, declination +04°338, but it is not easy to identify, since its apparent magnitude is only 9.5. Barnard found that it has a very rapid proper motion of 10.27 seconds of arc per year; this is the greatest known, so that the star tracks across a distance equal to the apparent diameter of the Moon in only 190 years. No other known star has an annual proper motion of as much as 5 seconds of arc. At its distance of 6 light years, Barnard's Star is also our nearest stellar neighbour apart from the three members of the Alpha Centauri group. It is a true celestial glow-worm, with a luminosity only 0.00045 that of the Sun. (Incidentally, it is now approaching us at a rate of 67 miles (108 km) per second, and will be at its closest to us in the year 11800. The range will be no more than 3.85 light years, but it will then start to draw away again, and there is no fear of an eventual collision.)

Using the Sproule telescope, van de Kamp began a long series of photographic measurements of Barnard's Star, hoping to detect an astrometric "wobble". He knew that it would not be very pronounced. For example, over a range of 10 light-years our Sun's "wobble", due to the effects of Jupiter, the most massive planet in the Solar System, would be less than 2 milli-arc seconds — and one milli-arc second is equal to the apparent thickness of a human hair seen from a distance of two miles (3 km). But by 1969 van de Kamp had taken over 3000 plates, made up of over 10,000 exposures on 766 nights. He found a distinct "wobble", and attributed this to a planet about 1½ times as massive as Jupiter, orbiting Barnard's Star at a distance of 400,000,000 miles (640,000,000 km) in a period of 25 years. In December 1970 he joined me in a BBC Television *Sky at Night* programme and described his results. "A body only 1½ times as massive as Jupiter cannot possibly be a star," he said, "and so it must be a planet." Later he made new measurements and announced the existence of two planets, each comparable in mass with Jupiter and with periods of 12 and 26 years respectively. "They must be very cold," he commented, "and extremely unpleasant places." No telescope available at that time could show them; the estimated magnitudes were around +30.

Unknown to van de Kamp, Nicholas Wagman, Director of the Allegheny Observatory, had also been making measurements, but with different results. Absolutely no "wobble" was detected. Next, George Gatewood and Heinrich Eichhorn, at Allegheny, used a telescope more powerful than van de Kamp's, again with negative results. Gradually the truth emerged. The apparent irregularities were due not to the star, but to problems with the Sproule telescope. Van de Kamp's results were spurious; the planets did not exist.

Next came the strange episode of VB8, another dim red dwarf in Ophiuchus, 21 light-years away. The star had been discovered by Georges van Biesbroeck, a Belgian astronomer who had worked at Heidelberg, Potsdam and Uccle before emigrating to America in 1915 and joining the staff of the Yerkes Observatory. Subsequently he had a highly distinguished career, and discovered two comets, 1925 VII and 1954 IV. However, he paid little attention to the obscure red dwarf, and nothing much was heard about it until 1983, when a team at the United States Naval Observatory, led by Robert Harrington, announced an astronomic "wobble" which could be due to an orbiting planet with a mass a few times that of Jupiter. In 1984 Donald McCarthy, from Arizona, used a new photographic technique to observe it directly. It was said to have a surface temperature of about 2224°F (1218°C), and McCarthy's paper claimed that "these observations may constitute the first direct detection of an extra-solar planet". Yet in 1985 French astronomers, using even more sensitive equipment, were totally unable to find the object, which by then had become generally known as VB8B. It had vanished as effectively as the hunter of the Snark, and was never seen again. Moreover, the tiny "wobbles" in the star itself proved to be non-existent. VB8B made an inglorious exit from the astronomical scene.

Alarm No. 3 involved a different sort of object, not an ordinary star but a pulsar with the catalogue number of PSR 1829-10. This caused a real furore, and led to a remarkable climax. First, however, what exactly is a pulsar?

Here we must go back to the way in which a star produces its energy. When the star is born inside a nebula, it begins to contract and heat up. Hydrogen must be the major constituent and when the inner temperature has reached a value of around 18,000,000°F (10,000,000°C), nuclear reactions begin; the nuclei of hydrogen atoms begin to band together to form nuclei of helium, with release of energy and loss of mass. (Our Sun is losing mass at the rate of 4 million tons (tonnes) per second, but please do not be alarmed — there is plenty left.) When the supply of available hydrogen is used up, the star must readjust itself and everything depends upon its total mass. In the case of a star like the Sun, the outer layers will expand and cool, while the core shrinks and heats up; the star passes through the red giant stage and then throws its outer layers away completely, leaving the core as a small, dense white dwarf which will, in the fullness of time, lose its light and heat to become cold and dead. However, a star which is much more massive than the Sun will have a much more violent fate. When the hydrogen fuel is exhausted, different reactions take over, but eventually there is a violent implosion, followed by an explosion. The star hurls most of its material away into space in what is termed a supernova outburst, while the core, now made up of neutrons, will be left on its own.

The neutron star may be no more than a few miles across, but its shattered parts of atoms are so closely packed that one could pack a thousand million tons (or tonnes) of neutron star material

The Crab Nebula, remnant of the supernova first observed by Chinese astronomers in July 1054. The Crab is 6000 light-years away and sends out radiation at most wave-lengths.

into an eggcup. There is a powerful magnetic field, and the neutron star spins quickly round, sending out pulsed radio emissions — hence the term "pulsar". The best known pulsar lies inside the Crab Nebula, known to be all that is left of a supernova seen in the year 1054, and which for a time became brilliant enough to be seen with the naked eye in broad daylight. The Crab pulsar spins round 30 times every second, but others are faster, with their spin periods measured in milliseconds.

Come now to PSR 1829-10. I first heard about it on 24 July 1991, when I was at Buenos Aires, in Argentina, for the 21st General Assembly of the International Astronomical Union, the controlling body of world astronomy. The IAU meets once in every three years, and this was its first foray into South America. On this occasion I had been asked to edit the official bulletin issued each morning to all the delegates, and I had commandeered part of an office in the main Conference building to act as an editorial headquarters. When I arrived there early in the morning of that particular day, I was greeted by Sir Francis Graham-Smith, the former Astronomer Royal. "We have some news about the new planet!"

I was taken aback. "What planet?"

"Discovered from Jodrell Bank by Andrew Lyne and his team. The planet's orbiting a pulsar, 30,000 light-years away."

I was frankly staggered. When a supernova explodes, it may send out as much energy as an entire galaxy, and a pulsar seemed to be the very last place where one would expect to find a planet, but I am not a radio astronomer, so I asked only two questions. "Have you an article ready?"

"Yes."

"Right, you've got the front page. By the way, what is the period of the planet?"

"Six months. That seems well established."

I admit that warning lights flashed in front of my eyes, and I made a quiet comment to my associate editor, John Mason, who was standing next to me: "That sounds like a reflection of the Earth's orbit round the Sun." But of course I did as I was asked, and the article duly appeared in the issue put out on July 25.

Normally, the pulses from a rotating neutron star can be timed with great precision. The period of PSR 1829-10 was 330 milliseconds, but Setnam Shemar, a graduate student working at Jodrell Bank with Andrew Lyne's team, found that it was not constant; sometimes the pulses arrived earlier than predicted, while at other times they were late. Shemar found that they were cyclic, and he decided that the cause of this strange phenomenon was movement. The neutron star was moving round the common centre of gravity of a binary system, and though the companion body could not be seen it was making its presence felt. When the neutron star was moving toward us (due allowance having been made for the overall motion in space) the pulses were early; six months later they were behind schedule, because the neutron star was then in the far part of its orbit and the pulses had further to travel. Everything pointed to the existence of a neutron star with 1.4 times the mass of the Sun, together with a planet 10 times as massive as the Earth, moving round the star at a distance of about 67,000,000 miles (108,000,000 km) — much the same as the distance between the Sun and Venus.

On 8 December 1991, back in London, I was joined in the television studio by Andrew Lyne, Setnam Shemar and the third member of the investigating team, Matthew Bailes. By then the reality of the planet was widely accepted, and I did no more than cautiously repeat my misgivings about the six-monthly period. Various suggestions were made with regard to the possible origin of the planet. Could it have been formed from a disk of material left round the neutron star at the end of the supernova outburst? Was it a late addition to the system, captured by the neutron star well after the end of the outbreak? Or could it be that the pulsar itself was not formed in the usual way, but was the result of a collision between two white dwarfs which merged, providing so much mass that the nuclear material was forced into the neutron stage? Opinions differed; nobody really knew, but subsequent papers went so far as to speculate about the possibility of life on this peculiar world.

Then, in January 1992, Lyne found the answer. There had been a slight error in the measured position of the pulsar, and when correcting for it the computer programme had made a disastrous simplification, assuming that the Earth's orbit round the Sun is a perfect circle, whereas in fact it is an ellipse. At once it became painfully clear that there was no need to assume the presence of any planet.

This was where Andrew Lyne showed his absolute integrity. It would have been so easy to blame instrumental error or computer failure, but he did nothing of the sort; he made an official statement explaining exactly what had happened, and accepting full responsibility for it. All in all, it is fair to say that he and his team emerged from the whole episode with greater credit than they would have done if the planet had really existed.

17

Poetic Moons

Listen, my children, and you shall hear
Of the only battle of Paul Revere...

These are the opening lines of a famous poem by Longfellow. It relates to an episode during the American War of Independence when, according to the poem, a garrison of colonists was saved from destruction by the bravery of Paul Revere who, alone and unaided, slipped through the British cordon to give due warning.

But who was Revere, and is the poem accurate?

Paul Revere was born in Boston on 1 January 1735. He became a silversmith, one of America's greatest, and for his artistic work alone he would be remembered, but he was also actively interested in politics and took a hand in the infamous Boston Tea Party in 1773. At the outbreak of hostilities he joined the army and by 1776 had risen to the rank of Lieutenant Colonel. Under his command the troops met with several reverses, and then came the disaster at Penobscot Bay, where a small force of pro-British colonists was holding out. A British force was sent to protect them and one of its officers, an 18-year-old lieutenant named John Moore, began to erect a fort on the shore, but before it was finished an American force, led by Revere, attacked. The result was decisive; the colonists were utterly routed. Only two ships survived, and the rest withdrew in disorder.

The British then prepared to march on Concord, where there was a munitions store. Revere decided that he must warn the Concord garrison that an attack was imminent, but how could this be done? According to the poem, he decided to row across the lake, close to the British ship HMS *Somerset*. This was possible on the night of April 17–18 because although the moon was full, it rose so far south that Revere could slip past unseen. This is what happened...

Or did it? Sadly, the truth seems to be rather different.

There were actually three riders: Revere himself, William Dawes and Samuel Prescott. Dawes and Prescott got through, Revere was captured, though subsequently released. Moreover, Revere was later charged with cowardice and incompetence, and though he was eventually cleared by a court-martial his military reputation was ruined. He died on 10 May 1818.

John Moore's career was rather different. He did achieve great distinction and, as Sir John Moore, was a commander during the Peninsular War against Napoleon. He was mortally wounded at the Battle of Corunna in 1809, though by that time the French had been repulsed. He was buried on 16 January, and here too there is a famous poem, this time by Charles Wolfe. It begins: "We buried him darkly at dead of night, the sods with our bayonets turning..." and the poem alludes to the pale light of the Moon. In fact the Moon was new on that particular night, and could not be seen at all. I suppose we must make due allowance for artistic licence.

Moore was, incidentally, my great-great-great-grandfather. I have to admit, however, that since he never married, my family tree is a little convoluted!

18

A Lightning Decision!

Free speech is still permissible in many countries of the world, but it was definitely not allowed in seventeenth-century Italy, and one man who found this out, to his cost, was the great scientist Galileo. He was finally exonerated, but not for several centuries after his death.

It was once believed that the Earth must lie at rest in the exact centre of the universe, with everything else rotating round it once in 24 hours. This was the view of Ptolemy of Alexandria, last of the famous astronomers of ancient times. Few people disagreed with him, but eventually doubts began to creep in and in 1543 a Polish churchman, Copernicus, published a book in which he claimed that the Earth moved round the Sun rather than vice versa. He met with furious opposition from the Church, who felt that any attempt to dethrone the Earth from its proud central position was heretical. The religious zealot Martin Luther thundered: "This fool seeks to overturn the whole art of astronomy. But as the Holy Scriptures show, Jehovah ordered the Earth, not the Sun, to stand still."

Copernicus himself avoided persecution by the wise decision to withhold publication of his book until the last days of his life. Galileo was less prudent. He was the first man to make systematic telescopic observations of the sky, from 1610, and he was an early convert to Copernicanism. He made no secret of his beliefs, and he never suffered fools gladly. (One Copernican, Giordano Bruno, was burned at the stake in Rome in 1600, partly (though not entirely) because he believed in a moving Earth.) So, in making his views known, Galileo was playing with fire — literally.

Initially he had friends in court in Rome, notably Cardinal Maffeo Barberini, but there were signs of approaching trouble in 1616 when the then Pope, Paul V, gave Galileo firm orders to

Nicolaus Copernicus (1473–1543), Polish cleric and astronomer.

stop preaching these dangerous doctrines. The doctrines in question were: (1) the Sun is the centre of the universe, and (2) the Earth is neither motionless nor the centre of the universe, but revolves round the Sun, as well as rotating on its axis once a day. Various books were then placed on the Papal Index, which meant that nobody was allowed to read them. Among these books was the original by Copernicus (*De Revolutionibus Orbium Coelestium*) which remained banned until 1835, shortly before Victoria became Queen of England. Galileo kept quiet, but in 1623 Cardinal Barberini became Pope, as Urban VIII, and this seemed to change the whole situation. Galileo began work on a new book, the English title of which was *Dialogue Concerning the Two Chief World Systems* — that is to say, the geocentric (Ptolemaic) and the heliocentric (Copernican). In 1630 he took it to Rome, showed it to the Holy Office, and asked permission to have it printed. Minor changes were made, and then Galileo was allowed to send the manuscript for publication. It appeared in February 1632, and at once the storm broke.

Unfortunately for himself, the Dialogues took the form of conversations between three imaginary people, one of whom, Simplicio, championed the old Ptolemaic theory, and whose arguments were very effectively demolished. It is widely (and probably correctly) believed that the Pope regarded Simplicio as a caricature of himself. Almost overnight, he changed from a

85

good friend into a remorseless enemy. In September 1632 Galileo was summoned to Rome, and in the following summer he was put on trial for heresy. Inevitably he was found guilty. Theoretically he could have been tortured; it does not seem that there was ever any real likelihood of this, but on 23 June Galileo was forced into a public recantation: "I do abjure, curse and detest the said errors and heresies... and I swear that in future I will never again say or assert, verbally or in writing, anything which might again give grounds for suspicion against me." He was then condemned to live alone at his villa in Arcetri, where he died on 8 January 1642. In fact the restrictions were relaxed after 1639, but Galileo had by then lost his sight. It was a sad end to the life of one of the world's greatest scientists.

There was one final piece of pettiness. It was proposed to erect a monument over Galileo's tomb, but the Pope expressly forbade anything of the kind.

The Ptolemaic theory lingered on for some time, but it is fair to say that after the middle of the seventeenth century few genuine scientists had much faith in it. Then, in 1687, Isaac Newton published his immortal *Principia*, and to all intents and purposes the idea of a central Earth was dead. Much later, a very fine observatory was set up in the Vatican. And yet so far as the Church was concerned, Galileo was still a convicted heretic.

The whole situation was ridiculous, and one man who seems to have realised this was John Paul II, the Polish Pope who was the reverse of bigoted and universally respected by Catholics and non-Catholics alike. Around 1979 the Pope set up a commission to re-investigate the whole Galileo case. The deliberations of the committee took thirteen years (!) but at last, in October 1992, a decision was made. Galileo had been right all along. The Earth really does move round the Sun and the spirit of Christianity is not undermined. So, in a major address about science and religion, the current Pope announced that Galileo had been exonerated.

Truth will out, but the interval between Galileo's condemnation and his rehabilitation amounted to over 359 years 11 months. At least nobody can accuse the Church of making hasty decisions!

19

The Strange Case of SS Lacertae

Have you ever heard of SS Lacertae? Probably not. It is an innocent-looking star of the tenth magnitude, living inside an equally innocent-looking open cluster, NGC 7209, in the little constellation of the Lizard. It is 3000 light-years away, and you can find the cluster easily enough with a small telescope, but there is nothing to draw special attention to SS Lacertae. Yet it has caused astronomers a great deal of heart-searching. It used to be an eclipsing binary; now, apparently, it is not.

Eclipsing binaries are made up of two components moving round their common centre of gravity. The classic case is Algol in Perseus, which normally shines as a star of the second magnitude. Every 2½ days it starts to fade, taking over four hours to drop to well below magnitude 3, remaining dim for twenty minutes and then recovering.

What is happening is that the main star is being periodically covered (or partly covered) by a less luminous companion. Another of these eclipsing systems is Beta Lyrae, close to the brilliant Vega; here the two components are less unequal, and there are alternating deep and shallow minima.

Telescopic eclipsing binaries are very common, and in 1921 the German astronomer Cuno Hoffmeister found that SS Lacertae was one. The period was 14 days 10 hours; the minima were slightly unequal, so that one component was rather more powerful than the other, and the fact that the secondary minimum did not fall exactly half-way between successive primary minima meant that the orbits of the two stars had to be somewhat eccentric. Both were normal hot stars of spectral type B9, with surface temperatures of over 10,000°F (5600°C); the masses were about three times that of the Sun.

Nothing unusual in all this. But during the years following

Hoffmeister's discovery, the eclipses seemed to become less marked, and by 1950 they had stopped altogether. Since then, SS Lacertae has shone with a steady, unchanging light. So what has happened? A star does not suddenly vanish; moreover, the star looks as bright now as it did when it was at maximum in the pre-1950 period.

It is tempting to suggest that the angle at which we see the system has changed. If we were viewing it from an edgewise-on position in 1950, we would see eclipses; if by now we were seeing it full on, we would not. Unfortunately this plausible explanation does not work. The stars would still be moving round their common centre of gravity, and this would cause regular changes in the positions of the dark lines in their spectra, due to the famous Doppler Effect*. Yet nothing of the sort is seen. There seems absolutely no escape from the conclusion that both the components are still there, but are no longer moving round each other.

It might be suggested that one of the pair suffered a violent outburst, which broke it free from the pull of its companion, but this would have shown on photographs taken from Earth, and again there would be changes in the spectrum. In any case, SS Lacertae is made up of stars which would not be expected to behave in such a way.

All in all, the most probable answer seems to be that the system was disrupted by the arrival of another star — and there are at least fifty stars in NGC 7209.

If so, then the intruder would have thrown the two components of SS Lacertae apart, so that they began to recede from each other. In this scenario, they are now almost lined-up as seen from Earth and moving further and further apart all the time.

It is therefore just possible that eventually they will become so widely separated that the Hubble Space Telescope (or its successor) will be able to see them individually.

* The Doppler Effect is the change in frequency of a light wave. With an approaching object the wavelength is shortened and the light-source appears "too blue". With a receding object the wave-length is increased and the source appears "too red".

Whether or not this explanation is correct remains to be seen. It sounds decidedly glib, but it is not easy to think of anything better. No other similar cases are known; in our experience SS Lacertae is unique.

20

How the Lunar Craters *Weren't* Formed

Much has been heard about the origin of the craters of the Moon. Nowadays most people believe that they are impact structures, though there are still a few supporters of the volcanic theory, according to which the craters are of the same type as our own calderae. But there have been other ideas too, so let us pause to look at a few of them.

There is, of course, the volcanic fountain theory, described by James Nasmyth and James Carpenter in their classic book published as long ago as 1876. They assumed that a volcano spewed out material which was deposited in a ring, building up the crater walls. Of course, this idea is not tenable, but neither is it at all outlandish, so I will say no more about it here. Neither will I dwell upon a theory by one F. Benario who, in 1953, conducted experiments in which he suspended pulverised plaster over soda water and Alka-Seltzer, producing craters in the plaster.

Let us turn instead to tidal theories, which date back at least to 1890, with papers by two Germans, H. Zehäder and H. Ebert. Here we go back to the time when the Moon was rotating rapidly on its axis. The Earth's pull produced tides in the magma, which poured out through openings in the lunar crust. At ebb tide the magma withdrew, leaving lava which immediately solidified. During the following high tide, the wave ejected the hardened masses vertically and gradually a crater was built up. The overall scheme was later developed by a Bulgarian, N. Boneff, who likened the action to that of a pump. He even maintained that in the far future the Moon will draw close to the Earth and cause similar craters here. However, we now know that both the Earth and Moon will be destroyed when the Sun changes it structure and becomes a red giant star, in roughly 5000 million years from now, long before any tidal forces could bring the Moon close to us again.

Next there was Amédée Fillias, a French engineer, who in 1961 proposed a theory involving the expansion of the Earth's core. The plastic crust was laminated and "lips" or circular ranges formed around the breaks; if the crust were completely perforated the bottom of the ring would break, producing a central peak. Against this we have a Norwegian, Ingolf Ruud, who put forward a core-contraction theory, so that when the crust was stretched beyond the breaking point craters were produced. Ruud's theory came out in 1934. I doubt if he and Fillias ever met; had they done so, their conversation would have been most interesting!

Let us now abandon every vestige of common sense. In 1846 a Captain Rozet, in France, suggested that temperature variations in the lunar crust would produce the equivalent of whirlwinds, ending up as craters. Even better was an Austrian, Josef Weisberger, who solved the whole problem very simply by denying that there were any lunar mountains or craters at all — they were simply storms and cyclones in a very dense lunar atmosphere. By 1936 Weisberger's thesis was complete, but no scientific journal would publish it, so he did so himself, with remarkable results. "For over two years I have, with complete abnegation and considerable financial sacrifices, attempted to bring authoritative astronomers to pronounce a scientific verdict upon my remonstrances," he wrote. "My treatises have not been answered; they have been returned or even ridiculed." He referred to statements "the inaccuracy of which is apparent even to the simplest mind", such as the widespread theory that the dark patches near the Moon's terminator are shadows cast by peaks or crater walls. I quote: "In the Moon's waning phase, when the 'shadows' are lengthening, the subsequent measurements will yield a progressive rise of the figures for the 'altitude' of the alleged 'craters' until finally the altitude found by the last measurements will be a multiple of that resulting from the first measurement of the narrow 'shadow' at the beginning of its existence." I hope that nobody will ask me exactly what he means…

A very early lunar photograph taken by L.M. Rutherford in 1868, together with some of James Nasmyth's plaster models which illustrate his attractive but wrong "volcanic fountain" theory of crater origin.

Weisberger was rather difficult to refute because, like the Flat Earthers and the astrologers, he threw overboard every ounce of conventional science.

The first ice theory seems to have been due to another Norwegian, Ericson, in 1886. Here we have a permanently cold Moon, so that the craters are nothing more nor less than frozen lakes. In 1914 a Spaniard, Rincon-Gallardo, went one better by suggesting that the lunar "seas" were the original lands and that the bright upland areas were frozen seas. In these frozen seas, volcanic eruptions caused local melting and craters resulted; rills were produced by contractions of the ice.

However, the ice theory was really worked out in 1889 by a tea-planter, E.E. Peal. This time it is the maria which are frozen water-lakes, and the entire Moon is coated with a layer of ice. Water vapour sent out from below the crust produces a localised, dome-shaped atmosphere, and this freezing material condenses round the volcanoes, forming icy craters. An icy Moon was also supported in 1925 by E.O. Fountain, of the British Astronomical Association, who believed that the Moon contains lighter materials than the Earth because its density is lower. Extensive oceans used to exist there, but as the atmosphere leaked away the water froze and was protected from the Sun's heat by a coating of meteoritic dust. Another supporter was F.J. Sellers, one-time President of the B.A.A., who also combined his ideas with the tidal picture. After the formation of an icy crust, hot water bubbled out through cracks at the time of high tide; it then froze and craters were built up.

However, the glaciation idea reached its climax in pre-war Germany. This was due mainly to an engineer named Hans Hörbiger (actually an Austrian), to whom practically everything was ice — even the stars which were ice-blocks in the sky rather than suns. Space, opined Hörbiger, is filled with rarefied hydrogen. The Sun, which is the most important body in the universe, is drawing all the planets in toward it because the planets are braked as they push their way through the hydrogen. Very small planets can drop into larger ones and the Moon will in time drop on to the Earth. (When an ice-planet hits the Sun, it produces a sunspot.)

At present the Moon is spiralling down, but it is not the first of its kind; there have been six previous moons, all of which have approached the Earth before disruption. Each time this happens there are major upheavals, one of which occurred around

The watchful dinosaur as visualised by Paul Doherty.

65,000,000 years ago and wiped out the dinosaurs.

To follow Hörbiger through: the penultimate Moon was at the peak of its career in near-historical times. The approach of so large a piece of ice lowered the Earth's temperature drastically, and terrified humanity had to migrate towards the equator. When the ice-Moon disintegrated, pieces of chilly material rained down on to the Earth; the Earth itself, which had been twisted out of shape, snapped back into a spherical form, with devastating earthquakes and storms, after which the equatorial waters flowed back to higher latitudes and, naturally, caused the Biblical Flood.

There followed a period of blissful calm. Then, between 13,000 and 14,000 years ago, our present Moon was captured and there were more earthquakes and eruptions, one of which submerged the island of Atlantis. The outlook for us is, I fear, not good — at least according to Hörbiger and his later followers.

In Hitler's time the ice theory was popular in Germany — it was called WEL — and at one stage the Government issued an edict to the effect that one could still be a good Nazi without accepting universal ice. A convert, albeit in modified form, was a well-known lunar observer, Philipp Fauth, whose books describe a Moon which has an icy crust hundreds of feet deep. Fauth was the author of a large, if rather inaccurate lunar map, and he had a considerable following, which for the time being gave the ice theory an aura of respectability. Unfortunately Fauth was not blessed with tact, and anyone who disagreed with him was automatically classed as a mortal enemy, which made normal argument rather difficult.

Fauth died during the war. WEL did not die with him, and Hörbiger Institutes lingered on for some time. No doubt Hörbiger himself would have been somewhat mortified to find that when Neil Armstrong first stepped out on to the Sea of Tranquillity he had no need to put on skates.

If ice is inadmissible, what about coral atolls? Here we come to Dr Edward Gates Davis, sometime President of the Astronomical Society of Kansas City, whose theory appeared in 1920. In the ancient seas of the Moon, coral built up; the accumulation of coral-bearing material from the large atolls, such as Tycho, explains the ray systems. Some craters were filled to the brim with sediments, so that central peaks could not be seen, while others were not. Davis made some intriguing calculations; thus in the crater Copernicus he assumed a coral growth rate of about 30 inches (76 cm) a year, so that the entire rampart could be completed in a mere 68,000 years. Lunar life was finally ended by violent crustal upheavals, but the coral atolls remain. Atolls were also supported by D.P. Beard in 1923, but as he added little to Davis' theories I think we may pass him by.

Without question, the king of all lunar theorists was Sixto Ocampo, of Spain, who was an engineer by profession. He eventually published his theory in 1949 in South America, because the Barcelona Academy of Arts and Sciences, to which he originally submitted it, turned it down. Ocampo went so far as to claim that an unscrupulous British astronomer had stolen his theory and was planning to publish it as original, thereby depriving Spain of the glory of the discovery.

Ocampo began his presentation by proving that the Moon does not rotate on its axis. Libration in longitude is due to the fact that the Earth-turned hemisphere is more massive than the far hemisphere, so that tidal effects keep it facing us (though how he reconciles this with non-rotation is decidedly unclear). The Moon used to be inhabited. There were two main races which, predictably, constructed atom bombs and proceeded to try to blow each other out of the Solar System. The bombs produced craters, until the whole of the Moon was pitted; the Alpine Valley and the Straight Wall are old engineering works. The last two craters produced were Tycho and Copernicus. Rays were caused by ejecta following the nuclear blasts, but the fact that the ray systems are different in form and texture, combined with the even more obvious fact that some craters have central peaks while others do not, show that the two sides used different types of bombs. The last great bangs on the Moon fired the lunar seas, which took off *en bloc* and fell back to Earth, causing — yes! — the Biblical Flood.

I doubt if anyone will ever surpass Ocampo, so let us allow him to have the last word. I will merely add that he died immediately after the publication of his thesis. His life's work was done.

21

The Lonely Brown Dwarf

A normal star is a gaseous globe, shining because of nuclear reactions going on deep inside it. In the case of the Sun, helium is being built up from hydrogen, with the release of energy and loss of mass. (In fact the Sun is losing mass at the rate of 4,000,000 tons (or tonnes) per second, so that it weighs a great deal less now that it did when you started reading this paragraph. But please do not panic, the Sun will not change dramatically for several thousands of millions of years yet.) A planet, of course, has no light of its own and there are no internal nuclear reactions. But are there any cosmical "missing links"?

To trigger off nuclear reactions, the temperature at the core of an embryo star must rise to around 18,000,000°F (10,000,000°C), so that it must have sufficient mass. The largest planet in our Solar System, Jupiter, may have a core temperature of up to 50,000°F (28,000°C), but this is not nearly enough; Jupiter has insufficient mass to qualify as a star. It is generally thought that any star which is shining by nuclear processes must have a mass 9 times that of Jupiter, which is about 8 per cent the mass of the Sun. If the initial mass is smaller than that no reactions will start. As the newly formed star condenses from the interstellar cloud, its core temperature will never reach the critical value and the star will simply shine feebly as it shrinks. It has become a Brown Dwarf, and will eventually lose all its light and heat, turning into a cold, dead globe.

Obviously, Brown Dwarfs are not easy to find, because they are so feeble and we cannot hope to locate any which are not reasonably close to us by stellar standards. A few have been identified with reasonable certainty, but these are either low-mass companions of normal stars or contained in clusters. But in 1997 a new Brown Dwarf was found, very much on its own in the Galaxy.

The story really began in 1987, when a Chilean astronomer, Maria Teresa Ruiz, decided to make a systematic search for stars known as White Dwarfs, which are very far advanced in their evolution and have used up all their nuclear "fuel". White Dwarfs are very dense, so that a ton of their material could be packed into an eggcup, but they do have appreciable inherent luminosity and they are very massive. The most famous White Dwarf is the dim companion of Sirius, which is as massive as the Sun but has a diameter of only about 25,000 miles (40,000 km) — smaller than a planet such as Neptune. Before becoming a White Dwarf, a star must pass through the red giant stage, as our own Sun will do in the far future.

For her hunt, Ruiz used the 39-inch (1-metre) Schmidt camera at the observatory of La Silla, in the Atacama Desert, where seeing conditions are superb. Her method was to photograph an area of the sky and then, some years later, take a second photograph of the same area. She looked for faint objects which had moved appreciably during the interval. Perceptible movement indicates relative closeness, and so objects of this kind were good White Dwarf candidates.

When checking plates of a region of the constellation of Hydra (the Watersnake), Ruiz detected a very dim star of magnitude 22.3 which had shifted quite markedly; it was moving at a rate of 0.35 of an arc second per year. At once she obtained a spectrum — and had a surprise. The object was not a White Dwarf at all; its spectrum was quite different, and moreover the star was red. At a range of about 33 light-years, this meant that the surface temperature was below 1700°F (927°C). The object could only be a Brown Dwarf, and Ruiz decided to give it a name; she called it Kelu-1, since "kelu" means "red" in the language of the Mapuche people, the original inhabitants of Central Chile.

What makes Kelu-1 so special is the fact that it is so isolated. As far as we know, there are no other stellar objects anywhere near it, so that it is lonely by any standards. Its estimated mass is about 6 per cent that of the Sun, which is equal to 75 Jupiters. It is a splendid example of a stellar missing link.

The NTT (New Technology Telescope) observatory at La Silla, Chile. The entire observatory rotates with the telescope.

What would it be like, viewed from close range? Of course the surface is too hot to support any form of life, and the chances of its being the centre of a planetary system seem to be very low indeed. From Kelu-1, the Sun would be visible as a star of about the fifth magnitude.

Kelu-1 will never become a normal, nuclear-burning star; it has too little mass. It will go on shining feebly for an immense period until it ends its inglorious career as a dead, lightless Black Dwarf, but this will not be for many thousands of millions of years yet. Perhaps, before then, some futuristic spacecraft will have passed by it and surveyed it. If so, the crew will certainly reflect that they are looking at one of Nature's failures — a star which never "made the grade".

22

Sister Marie — and Others

It seems that astronomy attracts more than its fair share of people who depart very markedly from the conventional line. These folk are often dismissed as cranks. Perhaps it would be kinder to call them independent thinkers. I am not here referring to the genuine cranks, notably the astrologers and the creationists, but to those who simply go off at a tangent. Let us take a brief look at some of them.

So far as I am concerned pride of place must go to Sofia Richmond, otherwise known as Sister Marie Gabriel. In fact, her full name is Sofia Segatis Paprocki Orvid Piciato. Apparently she was born in Novgrudok in Poland, in 1941, but came to England with her family in 1947 and has remained here. She went to school in Manchester, and it was while she was there that she decided to do her best to become a saint. Meantime she lives in Cricklewood, and it was from here that she hit the headlines in 1994 by taking full-page advertisements in some of the most prestigious daily newspapers, including *The Guardian*. A major crisis, she warned us all, loomed ahead.

It had already been announced that a comet — Shoemaker-Levy 9 — was to hit the planet Jupiter, and it actually did; the results were most interesting, though the long-term effects on Jupiter itself (or anything else) were precisely nil. However, Sister Marie Gabriel had her own ideas. First, she maintained that the wandering comet was none other than Halley's, which had last passed through perihelion in 1986 and which astronomers now knew to be a long, long way away, well beyond the orbit of the planet Uranus. Sister Marie did not explain how Halley's Comet had achieved an about turn, but she was convinced that it had done so and was *en route* for Jupiter. This, of course, would also have dire consequences for the Earth and the rest of the Solar

System. I can do no better than quote her own article, from *The Guardian* of 4 February 1992:

URGENT PUBLIC SAFETY ANNOUNCEMENT
World News Flash!!!
From Astronomer Sofia Richmond

HALLEY'S COMET IS ABOUT TO COLLIDE WITH THE PLANET JUPITER BY MID-JULY 1994 CAUSING THE BIGGEST COSMIC EXPLOSION IN THE HISTORY OF MANKIND.

A Warning Ultimatum from God to all Governments
SOS to the Pope!

BY 1st JULY ASTRONOMERS WILL CONFIRM THAT SOFIA (SISTER MARIE) WAS 100 PER CENT CORRECT IN IDENTIFYING THE COMET THAT IS HEADING FOR JUPITER AS HALLEY'S COMET.

...It is better to make a mistake on the side of maximum caution when facing the big unknown. After the comet hits Jupiter the planet could look like a second sun in the heavens... During a BBC interview at the Hilton Hotel Sofia (Sister Marie Gabriel) said that the true identity of the comet is Halley's Comet which Shoemaker and Levy saw in March 1993. The American astronomers sighted the broken fragments of Halley's Comet and renamed it after themselves.

Her message was sent personally to various world leaders, including Pope John Paul II, the Archbishop of Canterbury, John Major (at that time Prime Minister), President Yeltsin, HM the Queen, the Sultan of Brunei and, of course, Cliff Richard. On the day of the impact she advised that we should all wear dark glasses, stay in cold cellars or basements (to avoid the intense heat), close down all nuclear power stations, draw all the house curtains and avoid driving cars or travelling by bus or train.

Could we do anything to protect ourselves? Yes, of course. Sister Marie recommended that, for example, we should replace all beers and spirits with non-alcoholic beverages, abolish all crime, violence and obscenity on television, destroy all pornographic literature, and abolish all blood sports. I was, however, not quite sure why she also planned to reduce crime in Saudi Arabia by corporal and capital punishment...

Certainly she did not lack publicity. On television she was quite remarkable with ash-blonde hair tucked into a headscarf, a long cloak and dark glasses. She admitted that "perhaps I look a little younger than I really am".

The comet hit Jupiter on schedule; the Earth survived. But nothing daunted Sister Gabriel re-surfaced in 1995. On the front page of the national *Independent* for July 6 we read:

THE EXPLOSION OF HALLEY'S COMET
News flash!

The early return of Halley's Comet.
Astronomer (Polish, like Copernicus) predicts
HALLEY'S COMET HAS STARTED COMING BACK TO OUR WORLD!
UNEXPECTEDLY! NOW! (1st January 1992)
... *Scientists and astronomers seriously underestimated the colossal power and force of Halley's explosion in Feb/March 1991! The explosion dislodged Halley's Comet from its usual orbit, forcing it back to Earth now!*

Well, so far the comet has not returned, and astronomers do not expect it for another few decades yet; the next perihelion is due in 2061. Up to now nothing more has been heard from Sister Marie Gabriel, but no doubt she is simply biding her time.

Into the same category comes "the face on Mars".

In 1976 the Viking 2 spacecraft journeyed to Mars. It was made up of two parts: an orbiter and a lander. Once safely in Martian orbit, the two parts were separated; the lander came down, braked partly by parachute and partly by rocket power, while the Orbiter continued to circle the planet, acting as a relay. Before the separation, excellent photographs of the surface were taken and, in the region known as Cydonia, one rock gave an uncanny impression of a human face.

Astronomers pointed out that this was due merely to light and shadow effects, but the Independent Thinkers — were not convinced and became very vocal. Their main spokesman was one Richard Hoagland, from America, who has written a whole book

Halley's Comet, photographed in 1910 from the Helwan Observatory in Egypt. Exposure 7 minutes. In 1910 the comet was much more brilliant than it was at its last return in 1985.

on the subject, plus innumerable articles.

I quote:

"[The object] is a mile [1.6 km] long, 1500 feet [490 m] high. Around it, at distances of a mile or more, are five-sided pyramids. There are five of them. We can begin to see geometry and numbers repeating themselves. They are arranged in shapes that bring out the same mathematical relationships over and over again. The angles, the lengths, the directions are repeated. The conclusion must be that it was planned."

Inevitably there are accusations of a cover-up, and it is claimed that NASA is deliberately withholding vital information which would prove the existence of a Martian civilisation. It has suggested that Mars Observer, the American probe that went out of contact before arrival, was even sabotaged by the Martians....

Things were taken a step further in 1996 by a Briton, Peter Oakley, who lives in the south-west of England. He has been concentrating on the ancient stone circles in the village of Avebury. He has also visited the Pyramids, and has come to the astonishing conclusion that both these structures are linked with the Martian "face" in Cydonia. Again I quote:

"Cydonia appears to be an ancient city which has long since

been abandoned, and I think an intelligent alien life was responsible for its construction and its sister sites on Earth."* I would love to picture Martians busily at work in Wiltshire, but I have a feeling that Mr Oakley may have jumped the gun. Neither can I go along with Professor Courtney Brown of Emory University in Atlanta, who believes that because conditions on Mars have become hostile the Martians have established an underground colony in, of all places, New Mexico. "All the prestige I've got is resting on whether there is anything in this," he said. "I'd be crazy if I went public without being certain."† Well, he said it!

So far the Mexican Martians have remained well below ground level. Perhaps they will surface one day?

Another group of Independent Thinkers, led by a Mr Percy, maintains that no men have ever been to the Moon and that the whole Apollo story is a NASA hoax. The curious thing here is that they genuinely believe it, and are convinced of yet another cover-up. Presumably the idea came originally from a film, *Capricorn One*, about a faked journey to Mars — and in which NASA collaborated.

Finally, to politics. Just before the 1997 General Election a statement from the Monster Raving Loony Party announced that, if returned with an overall majority, they would shorten winter by the simple expedient of abolishing January and February. Of course, we must remember that the members of this party (to which I have the honour to belong) know quite well that they *are* loony.

Not so the members of the Newcastle branch of the official Green Party. With the solemnity of the truly barmy, they recently announced that in future they would time their meetings in accord with the phases of the Moon, meeting at new moon to discuss ideas and at full moon to act upon them. I feel that all one can say to this is: "No comment!"

* London *Daily Mail*, 5 March 1996
† London *Daily Express*, 22 August 1996

23

To Catch a Comet

If you are at all interested in ancient history, you have probably heard the name of Jean François Champollion. He was a French Egyptologist — and a good one. He was born in 1790 and, at the age of ten, read a paper to the Grenoble Academy in which he claimed that Coptic was the ancient language of Egypt. He moved to Paris and in 1809 was made professor of history at the Lyceum in Grenoble. In 1821 he achieved his first success in deciphering hieroglyphics, and went from strength to strength; all in all it is said that he was the real founder of scientific Egyptology. In March 1831 he received the chair of Egyptian antiquities at the Collège de France, a post which had been created specially for him. Unfortunately, he did not hold it for long; in 1832 he died while still at the height of his powers.

Now let us turn to the periodical comet Tempel 1, which was originally found on 3 April 1867 by William Tempel, from Marseilles; the discovery magnitude was 9. The orbit was worked out and the period found to be about five and a half years. It came back on schedule in 1873 and again in 1879, but in 1881 it passed within 52,000,000 miles (84,000,000 km) of Jupiter, and its orbit was drastically altered; the period was increased to 6½ years, and for many years astronomers lost track of it. In 1963 Brian Marsden, the well-known comet-orbit computer, undertook a detailed investigation and predicted returns for 1967 and 1972. A vague impression of what might have been the comet was later found on a plate exposed by Elizabeth Roemer, at the Catalina Observatory, in June 1967, but final confirmation had to wait until 1972. This time the maximum magnitude exceeded 11, and there was even a short tail. Since then the comet has been followed regularly, though it is always faint. The present orbit takes it from 138,000,000 miles (222,000,000 km) from the Sun out to 437,000,000 miles (704,000,000 km), well beyond the main asteroid belt; the period is 5.49 years, and the orbital inclination is 10.6 degrees.

What, you may ask, is the connection between Jean Champollion and Tempel's Comet? The answer is that if all goes well, a probe named in honour of the French Egyptologist will contact the comet at the end of the year 2005, touch down on its nucleus, and bring home a sample for analysis.

This will be the first attempt at a comet landing, though as we remember, Giotto went through the head of Halley's Comet in 1986 and sent back photographs from close range. Of course, a comet — large though it may be — has a very low mass by planetary standards, which is why they are so easily perturbed and forced into new orbits. A "landing" there will really be in the nature of a docking operation, because the gravitational pull of a comet is so slight. And, for that matter, it is by no means certain how "solid" a landing site will be. The nucleus of Halley's Comet proved to be coated with dark material, overlying a nucleus made up largely of ice mixed in with rubble; presumably Tempel's Comet will be rather similar. Yet, since a comet loses material every time it passes through perihelion (its closest approach to the Sun), short-period comets have largely wasted away, and Tempel will certainly be much less substantial than Halley.

The spacecraft carrying Champollion is to be known as Deep Space 4. The scheduled launch date is April 2003, in which case the spacecraft will enter a closed orbit round the comet in December 2005 and will remain circling its target for 140 days, during which time it will make a detailed map of the nucleus. Champollion will be detached and brought down. If the surface is soft and "mushy" (as may well be the case), Champollion will latch on by means of an anchor which will be rather like a harpoon. Four days will be spent on the comet, during which time samples will be collected and various other measurements made. There will be a drill to bore several inches into the surface. At the end of the four-day period Champollion will take off again and dock with Deep Space 4, transferring its samples to the capsule, which is scheduled to return home. Using its own rocket motors, the capsule will then begin the long return journey, finally landing back on Earth in May 2010 after an absence of over seven years.

It is certainly an ambitious programme, though it does not seem any more outlandish than packing a spacecraft with airbags and bouncing it down on the surface of Mars, as was done in 1997. Champollion was originally to have flown on a different spacecraft, Rosetta, which is due to reach another comet, Wirtanen, in 2012, but financial problems intervened here. Champollion is a NASA vehicle while Rosetta is the work of the European Space Agency, but the Americans pulled out because of the lack of money. Rosetta will now carry a small European lander, but no material will be brought home. Meanwhile Japan is entering the field; it is planned to obtain samples from an asteroid, Nereus, in 2006 and then concentrate on a comet, but at the moment it seems that Champollion will have the honour of being the first probe to bring us cometary material.

24

Curious Callisto

Jupiter, the Giant Planet, has an impressive retinue of satellites. Most of them are small, but four — the so-called Galileans — are of planetary size and three of them are larger than our Moon, while one (Ganymede) is actually larger than the planet Mercury. All four have been imaged by the various space probes and each has provided its quota of surprises. But to set the scene, here are their main characteristics.

Name	Mean distance from Jupiter (miles/km)	Orbital Period d. h. m.	Diameter (miles/km)	Density water=1	Magnitude
Io	262,000 (422,000)	1 18 28	2264 (3644)	3.55	5.0
Europa	417,000 (671,000)	3 13 14	1945 (3130)	3.04	5.3
Ganymede	666,000 (1,072,000)	7 3 43	3274 (5269)	1.93	4.6
Callisto	1,170,000 (1,883,000)	16 16 32	2981 (4797)	1.81	5.6

All four move in orbits which are practically circular, and also practically in the plane of Jupiter's equator, so that when seen telescopically they make a reasonably straight line (though, of course, it is not always that all four are on view simultaneously; one or more may be eclipsed, occulted or in transit, though it is seldom that all the Galileans are invisible at the same time). Ganymede is always the brightest of the four (really keen-sighted people can glimpse it with the naked eye) and Callisto is generally the faintest.

Ganymede and Callisto are icy and cratered, Europa icy and smooth, and Io red and violently volcanic, with sulphur eruptions going on all the time. (As it moves in the thick of the deadly radiation belts surrounding Jupiter, it must surely be the most lethal world in the entire Solar System.) Europa is quite different. There is little vertical relief, and almost no craters; it has long been suspected that below the icy crust there may be an ocean of liquid water. If so, it seems that the cause lies in the fact that Europa's orbit is not quite circular and the interior is being constantly churned by the changing gravitational pull of Jupiter, so that enough heat is generated to keep the water in liquid form. There have even been suggestions that this strange crustal ocean might support life, though frankly I take this with a very large pinch of cosmic salt.

When Europa crosses the lines of magnetic force in Jupiter's field, it disturbs the field noticeably, and if there is an extensive underground ocean this is only to be expected. Water, particularly salty water, is a good conductor of electricity, and if the Jovian magnetic field sets up a current in the water then this temporary magnetic field will make itself evident.

Jupiter's magnetic field is much the most powerful in the Solar System, and at times the magnetic tail can reach out as far as the orbit of Saturn. Io, of course, can also affect it, and a connection between the orbital position of Io and the radio emissions from Jupiter itself was established as long ago as the 1950s. But Callisto, outermost of the Galileans, was always assumed to be a totally inanimate sort of astronomical fossil, with a cold globe and a heavily cratered surface. Results from the Galileo spacecraft indicated that there was probably no definite core, but that the globe was made up chiefly of rock (around 60%) and ice (40%). There was absolutely no trace of any magnetic field, and no more than suspicions of an excessively tenuous atmosphere. To all appearances, nothing had happened on or in Callisto for thousands of millions of years. Therefore, it came as a tremendous shock when a team from the University of California found that Callisto was affecting the Jovian magnetic field in a way very like that of Europa.

How could this be? The investigators concluded that "The possibility of a liquid water ocean in Callisto is startling, but we have no other explanation for the near-surface, highly electrically conducting layer required by the observed induction signal." Perhaps "startling" is an understatement. "Inconceivable" would be more appropriate.

If it does indeed turn out that a subcrustal ocean exists in Callisto, then all our ideas about the Galilean satellites — and perhaps other bodies as well — will have to be drastically revised.

But there is another possibility. If we prove to be wrong about Callisto, then we may also be wrong about Europa.

There is another point, which is worth mentioning, though it refers to a date far in the future. No doubt there will eventually be manned missions to the Jovian system, but we will never be able to get too close; any astronaut venturing inside the main radiation zones will meet with a hasty and unpleasant death. But Callisto is more than 1,000,000 miles (1,600,000 km) from the Giant Planet, and is therefore, presumably, less hazardous than the rest, so that it may well be the site of the first landing. When this will happen we do not know, but certainly it will be a fascinating project.

Meanwhile, it is undoubtedly important to obtain more close range information from Callisto. It may, in the end, prove to be the most interesting, and also the most engimatical, of all the Galileans.

25

A Year on Icarus

The asteroids, or minor planets, were not always popular members of the Solar System. One irritated German astronomer went so far as to dub them "vermin of the skies", because their trails kept being found on photographic plates exposed for quite different reasons. Today, however, we recognise that they are full of interest. Among them are small bodies which swing well inside the main swarm and may pass close to the Earth. These NEAs, or Near Earth Asteroids, are divided into three classes:

1. **Amor asteroids**, whose orbits cross that of Mars but not that of the Earth.
2. **Apollo asteroids**, whose orbits do cross that of the Earth.
3. **Aten asteroids**, whose mean distances from the Sun are less than that of the Earth.

Several Apollo asteroids, including No 3200 (Phaethon) and 156 (Icarus) are distinguished by their close approaches to the Sun. Phaethon can skim by the solar globe at 13,000,000 miles (21,000,000 km), and Icarus at 17,000,000 miles (27,000,000 km). Icarus, discovered by Walter Baade in 1949, is a mere 1500 yards (1372 m) across, so that even when fairly close to us it remains faint. But in fact it is not a real "Earth-grazer", because its orbit is inclined at the high angle of 22 degrees, and there is no immediate fear of a collision. (The same is true of Phaethon.) The orbital period of Icarus is 409 days.

Obviously we can see no surface details on Icarus, though it has been contacted by radar; we know nothing about its topography. There is just a chance that it may be an ex-comet which has lost all its volatiles, though I admit I am sceptical about this. It spins quickly on its axis, and the Icarian "day" amounts to no more than 136 minutes.

Go to Icarus and you will have very little "weight". It would be childishly easy to throw a cricket ball clear away from the

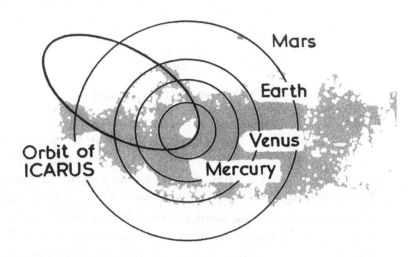

The orbit of Icarus. This was the first asteroid to be discovered passing inside the orbit of Mercury, but others have been found since.

little asteroid, and if you jumped up you would "float", taking a very long time indeed to come down. For a while you would turn yourself into a temporary asteroid in your own right. Bringing in a spacecraft would be more in the nature of a docking operation than a conventional landing. But just suppose we could go there and spend an Icarian year on the surface — what would we see?

Let us begin our journey at perihelion, 17,000,000 miles (27,000,000 km) from the Sun. The first thing that strikes us is the temperature. A thermometer shows about 900°F (almost 500°C), which is enough to make the rocks glow dull red. Heatproof clothing is an obvious necessity, and we are in a region which is dangerous in other ways too, because we are being bombarded by energetic particles and short-wave radiations emitted by the nearby Sun. The Sun itself is huge, covering an appreciable part of the sky and looking truly fearsome.

But because of Icarus' quick spin, it is not long before we are carried round to the night side. At once the temperature drops, though the rocks still retain enough heat to make them untouchable. Now the sky is black, and the stars shine out. We can see one brilliant planet, Mercury, though it still appears as a point of light rather than a disk. As we watch it seems that we can see the stars moving across the sky by virtue of Icarus' rapid rotation.

The landscape itself is subdued, because the alternate heating and cooling tends to level it, but here and there we can see craters produced by small bodies impacting the surface. With the return of daylight, the curvature of the globe is very evident; Icarus is only roughly spherical in any case. There are no major hills or valleys. Moving around means using the motors built into our space suits, because walking is out of the question — we are practically under conditions of zero gravity.

As Icarus begins to swing away from the Sun the temperature falls quite quickly. In a few weeks we have reached the distance of the orbit of Mercury and, if conditions are favourable, we may have a good close range view of it, so that we can make out the Mercurian craters and valleys. Next we reach the distance of the orbit of Venus, and that uninviting planet shines brilliantly during our brief period of night. By the time we have moved out to 93,000,000 miles (150,000,000 km) from the Sun things have changed dramatically; the rocks have ceased to glow, the Sun itself is reduced to the size we are used to at home, and the torrid heat is giving way to increasing cold.

Between six and seven months after our arrival, Icarus reaches its aphelion, at 183,000,000 miles (295,000,000 km) from the Sun. The path of Mars has been left behind, and the temperature has fallen far below zero; it is strange to reflect that a little while ago the rocks were so amazingly hot. Earth is a brightish speck in the distance, but now we can see some of the main-belt asteroids: Juno, Victoria, Iris, Flora and Phocaea. They are all much larger than Icarus, and all are moving near the inner edge of the main swarm, though the four giants of the asteroid zone (Ceres, Pallas, Vesta and Hygeia) are considerably further out. Binoculars will show many of the main-belt asteroids, and far away Jupiter looks rather brighter than it does from Earth, even from its range of over 200,000,000 miles (322,000,000 km).

Aphelion past, we start to swing inward again. This time we make a fairly close approach to Mars, and we can make out the two Martian satellites, Phobos and Deimos, which may themselves be ex-asteroids that were captured by the Red Planet long ago. Yet even Deimos is over nine miles across, so that Icarus is a true midget by any standards.

Slowly at first, then more rapidly, the temperature rises. Our thermometer creeps up and the rocks begin to glow. The Sun swells alarmingly, and at last after 409 Earth days, over 4400 Icarian days or one Icarian year, we are back at perihelion. Once more the rocks are bright red and remain uncomfortably warm even when carried across into the night regions. Icarus is in the grip of its long, torrid "summer". It is time for us to leave....

Will anyone go to Icarus in the future? It does not seem very probable. There may be a brief expedition, but not when Icarus is near perihelion, and any visiting astronauts will not stay for long. This strange, unwelcoming little asteroid must surely rank as the Devil's Island of the Solar System.

26

The Lighter Side of Space

Space research is a serious business. There have been many hitches and a few tragedies. But now and then an element of farce creeps in so, for a moment, let us look back at a few decidedly hilarious episodes.

The first concerns Hermann Oberth, the Romanian who in 1923 wrote the first properly scientific book about space research. It sparked off a wave of interest which led in time to men on the Moon. I met him on quite a number of occasions, but we were never able to talk directly because, while he spoke German and Romanian, I speak English and French. Our relations were very cordial so far as they went, which under the circumstances could not be very far. However, the curious event which concerns us here took place in 1929, when I was a boy of six.

Oberth's book had made him famous, and when the eminent film-maker Fritz Lang decided to put space travel on to the silver screen it was to Oberth that he turned for advice. Oberth, then teaching at a school in Mediash in Romania, was happy to oblige, and he travelled to Berlin eager to take part. He was not concerned with the script; that had been written by Mrs Fritz Lang, and Oberth was to attend to the technical details. Then Lang had another idea. Why not build an actual Oberth rocket and launch it on the day of the film première?

This would certainly have made for good publicity, but first the rocket had to be built and Lang had overlooked several things. The most important of these was that Oberth, brilliant theorist though he may have been, was not a practical engineer. and he had absolutely no experience in organisation. He was unused to life in a big city, and before long he found himself decidedly out of his depth. Neither did he choose his assistants wisely. The first was Rudolf Nebel, who had been earning a living by selling mechanical kitchen gadgets. The second was Alexander

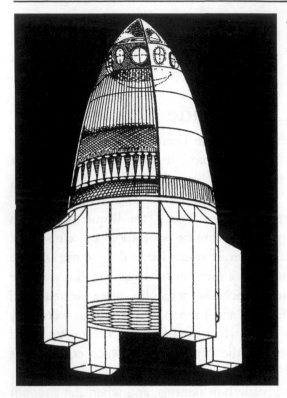

A diagram of the Oberth rocket — which never flew!

Shershevsky, a Bolshevik who, for various reasons, was reluctant to return to his native Russia. It was not a promising beginning.

Oberth's book had stressed the need for liquid fuels rather than solids such as gunpowder. Of course this is correct, but one critic had claimed that nothing of the kind could ever work, because as soon as liquid oxygen was combined with another ˆuel, such as petrol, the result would be an explosion rather than a process of steady combustion. Oberth decided to make a test. He poured liquid air into an open bowl — not liquid oxygen, which was regarded as too dangerous — and shot a thin stream of petrol inward. Not surprisingly, the mixture went off bang, with disastrous results to the windows of the improvised laboratory. Later tests went rather better, but at one stage there was a much more violent explosion in which Oberth was injured, fortunately not seriously, but he was badly shaken and time was running short. The film première was due in a few weeks ...

There was also the problem of a launching site. The first choice was a small, flat island in the Baltic, named the Griefswalder Oie. The local authorities were not impressed with the idea, and pointed out that on the Oie there was an important lighthouse which might, they maintained, be damaged by debris from falling rockets. In fact, it would have been difficult to damage the lighthouse even by a direct hit from a charging tank, but the Oie site was abandoned and the seaside resort of Horst was substituted. Nobody bothered to check that the Griefswalder Oie and its lighthouse were still well within the probable area of descent.

The film première was drawing nearer and nearer, and the rocket had not even been started. In an effort to speed things up Oberth changed his plans, but none of the experiments really worked, and it became painfully clear that the deadline was not going to be met. Oberth, rapidly approaching a nervous breakdown, left Berlin and hastened back to Mediash. The film company issued a statement to the effect that the launch had been postponed indefinitely; the film was released on 15 October 1929, but the Oberth rocket never flew at all.

Our next episode again involved Rudolf Nebel. It became known as the Magdeburg Experiment, and involved the bizarre theory of a hollow Earth.

The story really began as far back as 1869, with a vision announced by an American named Cyrus Teed. Even earlier there had been suggestions that the world might be hollow, and Teed adopted this theory hook, line and sinker. The whole universe, he said, was made up of the shell of the Earth, which is a hundred miles thick. We live in the interior, and beyond the shell there is absolutely nothing. We cannot see Australia above us because of the density of the atmosphere, so that there is no hope of pointing a telescope upward and watching kangaroos prance around. The Sun itself is half-bright and half-dark, and we cannot see it at all. Instead we have a sort of ersatz Sun, which is the reflection of the central one. When the Sun turns its dark side toward us, we experience night. There is no need to bother about the Moon, which is simply the reflection of the

Earth. Teed's book came out under the pseudonym of Koreshan, which is Hebrew for Cyrus, and the theory became known as Koreshanity Cosmology.

Pass on now to 1933. By then the early rocketeers, including Wernher von Braun, had set up the Rocket Flying Field near Berlin, and were busy with their pioneer experiments. These came to the attention of a Herr Mengering, who was a member of the Magdeburg City Council and a keen supporter of Koreshan. Very well then, would it be possible to launch a rocket vertically so that it would land somewhere in the region of Australia or New Zealand? Mengering believed so, and he made haste to contact Rudolf Nebel at the Rocket Flying Field.

Nebel's qualifications were dubious in the extreme, and later on his support for Nazism damaged his reputation even further, but he did see that there might be a chance of obtaining money to finance the building of a really powerful rocket. Mengering persuaded the City Council to make a substantial grant, and work went ahead. One proposed rocket was to carry a man to a height of almost a mile (1.6 km), and was to be a true monster as tall as a three-storey building. After reaching peak altitude the pilot was to eject by parachute and glide back to terra firma. A smaller rocket was to attempt the perilous journey to Australia. Needless to the say, the manned vehicle was never even started, but the smaller ones were, and by 1933 all was ready for the great test. It took place on June 9, but, alas, the rocket did not depart for the Antipodes; it rose no further than the top of the launching rack, after which it sank back to its starting position and refused to budge. A second attempt on June 13 was slightly more successful, since the peak altitude was no less than six feet (1.8 m). The supreme moment came on June 29, when the rocket really did clear its launching paraphernalia. The trouble was that instead of being vertical the trajectory was horizontal, and the journey ended in a sort of belly flop at a distance of roughly a thousand feet.

That was the end of the Madgeburg Experiment but, amazingly, it was not the end of Koreshanity Cosmology, which lingered on for some time after the end of the Second World War.

A space walk, 1920s vintage, from the film Frau im Mond.

The magazine supporting it, aptly entitled *The Flaming Sword*, made its final appearance in 1948.

Even later came the German Society for Geocosmical Research, based in Garmisch-Partenkirchen and led by Helmut K. Schmidt and P.A. Müller-Murnau. They differed from Koresh inasmuch as their hollow Earth extends beneath our feet infinitely in all directions. The Sun itself does not move, or even spin, but the inside of the Earth turns round, completing one revolution in 24 hours and giving us alternate day and night. Stars are spots etched on a crystal sphere surrounding the Sun. Australia cannot be seen overhead because the sky is too bright, and also because light does not travel in straight lines; it follows the curvature of the inside of the Earth. The last paragraph of Herr Müller-Murnau's book, published in 1957, states that "Mankind can only gain if it recognises that for good or bad we are living in an enclosed space."

All this contrasted sharply with the ideas of the International

Flat Earth Society, whose members maintain that the world is shaped like a pancake, with the North Pole in the middle and a wall of ice all round. (You can't go to the South Pole, because there isn't one.) Rather maliciously, I managed to put the Flat Earthers in touch with the Hollow Globers. For some reason or other they could not quite see each other's point of view, and to the best of my knowledge they are still fighting it out.

There have been many unconventional suggestions about space travel (quite apart from the UFOlogists, who are in a class of their own). I particularly liked the thesis of a Mr Theodore B. Dufur, of the United States, who in 1955 put forward the idea of an edible spaceship. It was to be made up of substances such as frozen oleo margarine, and was to be sent to the Moon. On landing you settle down in a convenient crater, drop a lid over the crater to keep yourself warm, and then nibble away at your spaceship until a relief expedition arrives. Mr Dufur's paper was even published in a prestigious American aviation magazine. Nobody has ever quite found out how it managed to slip through the editorial net.

Finally, let us turn to a different sort of national space programme, dating back to 1964. This time it comes from Zambia. I quote from a report published on 3 November 1964.*

"America and Russia may lose the race to the Moon, according to Edward Mukaka Nkoloso, Director-General of the Zambia National Academy of Space Research. His ten Zambian astronauts and a seventeen-year-old African girl are poised for the countdown. He said:

"I'll have my first Zambian astronaut on the Moon by 1965. My spacemen are ready, but we're having a few difficulties…we are using my own system, derived from the catapult." He went on: "To really get going we need about seven hundred million pounds. It sounds a lot of money, but imagine the prestige value it would earn for Zambia! But I've had trouble with my space-

---

* I described this in a somewhat flippant book, *Can You Speak Venusian?*, published in 1972, but I hope you will forgive my repeating it here. It is really too good to miss.

men and spacewomen. They won't concentrate on spaceflight; there's too much lovemaking when they should be studying the Moon. Matha Mwamba, the seventeen-year-old girl who has been chosen to be the first woman on Mars, has also to feed her ten cats, who will be her companions on the long space flight. I'm getting them acclimatised to space travel by placing them in my space capsule every day. It's a 40-gallon (180-litre) oil drum in which they sit, and I then roll them down a hill. This gives them the feeling of rushing through space. I also make them swing from the end of a long rope. When they reach the highest point, I cut the rope — this produces the feeling of free fall."

Sadly, the United Nations authorities turned down the Z.N.A.S.R. request for £700 million, but on 29 December 1968 Mr Nkoloso was in the news again. He congratulated the Apollo 8 team, who had made the first circumlunar flight, and then went on: "Let us make a Zambian rocket today. We shall never be content to remain behind other races. This is our heavenly destiny, our natural ambition and cultural hegemony." Meantime the astronauts remained in full training — cats and all.

I last heard of Mr Nkoloso in August 1971, when he announced that the Zambian Space Research Academy was building a new telescope "to see the planets clearly". The telescope was under construction at Chunga Valley, ten miles west of Lusaka, and two technicians and three engineers were working on it. It was to be officially opened in September with a display by twelve Zambian astronauts.

I was intrigued, and wrote to see whether I could secure an invitation to the grand opening. However, I had a letter from the Ministry of Technology in Lusaka saying that the opening had been postponed, and I never heard any more.

Oh well, it takes all sorts to make a world!

27

Alshain

Occasionally I find it entertaining to select a star — any star — and take a careful look at it. Of course, some stars are favourite objects of study, but there are many others which appear more or less nondescript. Just for a few moments then let us concentrate on the star Alshain, officially referred to by astronomers as Beta Aquilae.

Alshain is one of the stars making up the main pattern of the constellation Aquila, the Eagle. The leader of Aquila is the brilliant Altair, the twelfth brightest star in the entire sky. It is one of three stars making up what is now usually called the "Summer Triangle" — a nickname I casually used in a television programme almost forty years ago and which has now been adopted everywhere. The other two stars of the Summer Triangle

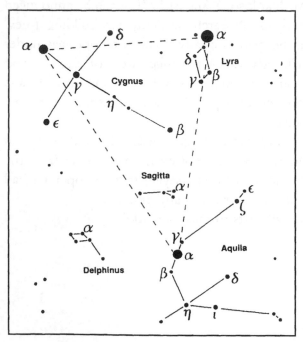

The Summer Triangle — Vega, Deneb and Altair.

are Vega and Deneb. The three are thus in different constellations — Vega is in Lyra, Deneb in Cygnus — and there is no
connection between them. For example, Altair is a near neighbour, only 10 times as luminous as the Sun, while Deneb is im-
mensely remote and could match 70,000 Suns. I used the nickname because the three stars are so prominent in British skies
during summer evenings, though of course this does not apply
to countries such as New Zealand and South Africa, where the
three are at their best during winter. (Moreover, Deneb is so far
north in the sky that from the southernmost tip of New Zealand
it never rises at all.)

Alshain has a declination of 6°24' north, so that it is so close
to the equator that it can be seen from every inhabited country.
It is also very easy to identify. Altair is flanked to one side by
Alshain, and to the other by Gamma Aquilae or Tarazed, so that
the line of three cannot be mistaken. Tarazed, of magnitude
2.7, is much brighter than Alshain, whose magnitude is only 3.7,
and this raises a rather interesting little problem.

In 1603 Johann Bayer compiled a new star map and allotted
Greek letters to the stars in each constellation. The second
letter of the Greek alphabet is Beta, and so Alshain should be
the second brightest star in Aquila, but it is not — it comes
eighth. It is surpassed by Alpha (Altair), Gamma (Tarazed), Zeta,
Theta, Delta, Lambda and Eta in that order (though Eta is a
Cepheid variable, and at minimum falls to below the fourth magnitude). This, in itself, would not be significant, because the
Greek sequence is often out of order, but it is curious to find
that older maps rank both Alshain and Tarazed as being of the
third magnitude. This would mean that they would be equal,
and the "line of three" would be symmetrical, which is certainly
not the case today. In a book published long ago, the French
astronomer Camille Flammarion gave the values as listed in old
catalogues. (See table on next page.)

Flammarion is quite definite: "Beta is actually greatly inferior
to Gamma, while up to the time of Tycho and Bayer both were
of the third magnitude. As it is not its position in the sky
which has influenced its denomination, it is certain that it has

	Alshain	Tarazed
130 BC, Hipparchus	3	3
960, Al-Sufi	3-4	3
1430, Ulugh Beigh	3	3
1590, Tycho	3	3
1603, Bayer	3	3
1660, Hevelius	4	3
1700, Flamsteed	3½	3
1800, Piazzi	3-4	3
1840, Argelander	4	3
1860, Heis	4	3
1880, Flammarion	4.0	3.3
(1998, modern	3.71	2.72)

diminished in brightness." "Denomination" refers to right ascension. In some cases Bayer gave Greek letters in order of R.A. rather than brightness, but even this could not justify Alshain's claim to the second letter of the alphabet.

There are various cases of stars which are alleged to have brightened or faded since ancient times. It has been claimed that Megrez, now much the faintest of the seven stars making up the main pattern of the Great Bear or Plough, used to be equal to the others, and that Castor used to be brighter than its "twin" Pollux, rather than a magnitude fainter. But we must always be cautious about claims of this sort, and generally the evidence is very slender. It is so, I feel, with Alshain, mainly because it is not the sort of star which would be expected to show secular change. Its spectral type is G8, which means that its surface is rather cooler than that of the Sun: around 10,000°F (5000°C). It is 4½ times as luminous as the Sun and is a mere 36 light-years from us, which by cosmical standards is close. It appears to be in a stable condition and there is no current evidence of any variation at all. All the same, look at it on any clear night and you will see the obvious difference between it and Tarazed. Moreover Tarazed is of spectral type K3 and is strongly orange in hue, whereas Alshain is only very slightly yellowish — in fact, most people will see it as white.

Since Alshain is cooler than the Sun, but also much more powerful, it must be larger, with a diameter of several millions of

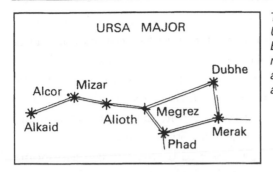

The seven main stars of Ursa Major (the Great Bear). Megrez is around a magnitude fainter than any of the others. Was it always so?

miles instead of only 865,000 miles (1,392,000 km) as with the Sun. Of course, telescopes show it as nothing more than a speck of light and no surface features can be seen on it, but probably it would show star spots if only we could see it from close range.

Does Alshain have a system of planets? This we do not know, but there is no harm in speculating, so let us assume that we are able to observe from a planet moving round the star, sufficiently far out to allow for a tolerable surface temperature. What would the sky be like?

During daytime on the planet there would, of course, be a very powerful yellow sun, probably active. At night the constellation patterns would be different from ours, because we would be observing from a different vantage point in the Galaxy. Our Sun would be rather dim and would shine as a star of rather below the fifth magnitude, so that it would be on the fringe of naked-eye visibility (assuming, of course, that our hypothetical astronomer has eyesight comparable with ours). No telescope of the sort we can make at the moment would show the Earth, or even Jupiter, but positional measurements carried out over a sufficiently long period would show that the Sun is "wobbling" slightly because of the gravitational pull of Jupiter. Remember, Jupiter is more massive than all the other planets in the Solar System combined, and moreover a radio astronomer on a planet moving round Alshain would be able to detect "radio noise" because of our transmissions. In the year 1999, according to our calendar he was, in theory, picking up programmes broadcast in 1963, when my own *Sky at Night* television series was already six years old....

But there would be something else, too. Alshain is not an isolated star. It has a dim companion, and this companion can be seen from Earth. I looked at it very recently with the 15-inch telescope at my observatory in Selsey, and it was plain enough. It is of magnitude 11.6, and is almost 13 seconds of arc away from the bright star, so that it is not "drowned" in Alshain's light. The position angle is about 150 degrees.

The companion was first reported by the Russian astronomer Struve in 1852, and was included in his double star catalogue as OE532. It is a red dwarf, of spectral type dM3, and its luminosity is only about 1/300 that of the Sun. It and Alshain share a common motion through space, almost half a second of arc per year, so that they are certainly associated and presumably make up a binary pair. But since they are so far apart, the orbital period round their common centre of gravity must be very long, amounting to at least 2000 years and possibly more. From our hypothetical planet it would appear as a reasonably bright star, but certainly not a "second sun". Its real distance from Alshain is not known accurately, but may be of the order of 160 astronomical units — that is to say, around 15 thousand million miles (24 thousand million km).

At the moment Alshain is approaching us at the rate of 24 miles per second, but this will continue indefinitely, and there is no fear of a collision in the far future. We have nothing to fear from this rather ordinary star in the celestial Eagle.

28

Names in the Sky

Would you like to have a star named after you, or one of your friends? Nothing is easier, say some well publicised "star registries". Pay us a suitable sum (say £50 or even £100) and we will arrange it for you, so that you will be remembered in perpetuity.

In fact things are not nearly so simple as this and a great many people have been victims of a particularly unpleasant confidence trick. The situation was highlighted in September 1996, when even John Major, then British Prime Minister, was lured into the trap. But first let us take a brief look at the various systems of nomenclature now in use.

(a) **Planets**. The five naked-eye planets, known since antiquity, were named for mythological gods — Mars, Jupiter and so on. The names are well selected and, for instance, the largest of the planets is named in honour of the ruler of Olympus, while the Red Planet is called after the god of war. The names were originally Greek, but we use the Latin equivalents. Ares, for example, is known to astronomers as Mars, though what might be miscalled "Martian geography" is officially termed "areography".

Planets discovered in telescopic times were named according to the same system: Uranus, Neptune and Pluto. Uranus and Neptune were accepted by tacit agreement; "Pluto", in 1930, was the suggestion of an eleven-year-old Oxford girl, Venetia Burney (now Mrs Phair), and was approved by the controlling body of world astronomy, the International Astronomical Union.

(b) **Asteroids.** The first asteroid was discovered in 1801, and the discoverer, G. Piazzi, named it Ceres in honour of the patron goddess of Sicily, where Piazzi lived. Asteroids discovered later were also given mythological names, but there are so many asteroids that the supply of gods and goddesses soon became exhausted, and different types of names were used, some of them decidedly bizarre. For instance Asteroid No. 2309 has been named Mr Spock, after a ginger cat which was itself named after the sharp-eared Vulcanian astronaut of the science fiction television series *Star Trek*. The discoverer of an asteroid is entitled to choose a name, but this has to be ratified by the nomenclature committee of the I.A.U.* Occasionally there are embarrassing mistakes. One asteroid was named 'Anvlad' after a Russian named Vladimirov, but the named had to be changed when Mr Vladimirov was exposed as a financial swindler!

(c) **Planetary satellites**. Again the early names were mythological, so that, for example, Saturn's family includes Titan, Iapetus and Rhea, all of which are associated in some way with the god Saturn. Again, the names given in modern times have to be ratified by the I.A.U. There is only one departure from the usual system: the satellites of Uranus were given Shakespearean names (Titania, Oberon, Ariel) though one came from a poem by Pope. I have always regarded this as most unfortunate, but it certainly will not be altered now.

(d) **Lunar features**. Several systems of nomenclature were used when telescopes first came upon the scene, but these were superseded by that of the Jesuit astronomer G. Riccioli, who drew a lunar map in 1651 and named the craters after famous people, usually astronomers. Riccioli's system has been followed, and of course extended to craters which were either too small for him to see or else were inaccessible. Remember, from Earth there is a part of the Moon which we can never see because it is always turned away from us.

(e) **Surface features on other planets**. With Mars, the lead was taken by R.A. Proctor, who published a map of the planet in 1867 and allotted names in honour of famous observers of Mars; we had Cassini Land, Mädler Continent,

*I cannot resist adding that Asteroid 2602, discovered from the Lowell Observatory in Arizona, was named "Moore" after me. I can assure you that I had nothing to do with the selection, and when told about it I was frankly taken aback.

Arago Strait and so on. (I always liked Beer Continent, honouring Wilhelm Beer of Berlin.) In 1877 a new system, based on Classical names, was introduced by G.V. Schiaparelli, so that, for example, Mädler Continent became "Chryse", while Cassini Land became "Ausonia". Schiaparelli's basic system is still followed, though it has been drastically extended and modified following the space research results.

Surface features on Mercury, revealed by the Mariner 10 space-probe, have been named after famous people including writers, artists and musicians, though there are some exceptions. For example, valleys have been named after radar installations, and scarps after famous ships of exploration and discovery. On Venus, the names of the features mapped by radar are invariably feminine, though James Clerk Maxwell, the Scottish physicist, had already been honoured with a mountain range before this edict came into force. Obviously there can be no permanent names for the transient features of the giant planets, whose surfaces are gaseous and always changing, and a nomenclature for Pluto has yet to be worked out. (I have proposed a scheme involving Underworld deities.)

Features on planetary satellites have also been allotted. For example, those on Saturn's satellite Dione are based on Roman mythology and history (Romulus, Remus, Aeneas, etc.) and those on Jupiter's satellite Callisto come from Norse mythology (Valhalla, Asgard, Grimr, Skoll).

(f) **Comets.** These have catalogue numbers, of course, but are also named after their discoverers, or (occasionally) after the computers of their orbits (the classic example being that of Halley's Comet). If there are two independent discoverers, both are included (Hale-Bopp, the bright comet of 1997. It was found first by Alan Hale and then, a few hours later, by Thomas Bopp). Sometimes the names are decidedly tongue twisting, for example Schwassmann-Wachmann, Honda-Mrkos-Padjusaková, and Churyumov-Gerasimenko.

(g) **The Stars.** Here we come to the unofficial "registries" which have been given so much publicity in recent years.

The constellation patterns which we use are essentially Greek though, of course, the original constellations have been modified and extended, and new groups have been added (notably in the far south of the sky, never visible from the Mediterranean area). In 1603 a German astronomer, Johann Bayer, compiled a new star atlas, and in each constellation allotted the stars Greek letters, in theory beginning with the brightest star, Alpha, and working through to Omega, the last letter of the Greek alphabet. Thus the brightest star in Lupus (the Wolf) is Alpha Lupi, the second brightest Beta Lupi, and so on. In fact, the strict alphabetical order is often not followed; thus in Gemini, Beta (Pollux) is brighter than Alpha (Castor), and in Sagittarius, the Archer, the brightest stars are Epsilon and Sigma, with Alpha and Beta very much in the "also ran" class. Obviously the system is limited, and there are many others in use. John Flamsteed, the first Astronomer Royal, was installed at Greenwich Observatory, by order of King Charles II, to compile a new catalogue. He did so, giving each star a number; thus in Orion, Alpha (Betelgeux) is also 58 Orionis, Beta (Rigel) is also 19 Orionis, and so on. Fainter stars are usually catalogued by their right ascensions and declinations — that is to say, by their positions in the sky.

Many of the naked-eye stars have also been given individual or proper names. A few are Greek (notably Sirius, from the old Greek word meaning "scorching") but most are Arabic, and come from the Arabs of around a thousand years ago. But in general, proper names are used only for the very brightest stars in the sky — those of the first magnitude — and a few special cases, such as Mizar, the double star in the Great Bear, and the variable Mira, in Cetus (the Whale). The rest are simply forgotten. I am quite sure that few astronomers would know that the obscure star Beta Piscis Australis, which is well below the fourth magnitude, has an old individual name, Fum el Samakah.

Bear one vital vital point in mind. *No stars are now given proper names*, and all names given to objects such as asteroids, comets and craters on planetary surfaces must be sanctioned by the Nomenclature Committee of the International Astronomical Union. This Committee has at present fifteen members, of which I happen to be one. Our recommendations are presented at the General Assembly of the Union, which is held every three years, and are then ratified so that they become official. So where does this leave the much publicised star registries? The answer is — nowhere.

The first agency seems to have come from America, and was followed by others. It was claimed that on payment of a set sum, a star could be given a chosen name. Innocent people took this at face value, and began sending in hard-earned cash. The agency advertisements were skilful by any standards, and before long the craze spread. Other agencies sprang up and were also established in Britain and elsewhere.

In fact, the names meant nothing at all. The buyers received charts indicating the position of the relevant star, and presumably each agency kept some sort of list though, of course, a different registry might well sell the same star to somebody else. Not that this mattered in the least —nobody would ever see the lists.

Initially this might have been interpreted as a piece of fun. (One enterprising gentleman in Japan has even persuaded some people to buy fishing rights in the lunar seas!) But, in fact, it soon developed a very unpleasant aspect. Bereaved parents were tricked into buying stars for their dead children. For example, not long ago I received a letter from a man living in Wales: "My daughter was aged four and last month she was knocked down by a car and killed. I am told that I can pay £70 and name a star after her. It sounds a wonderful gesture." Luckily he received my reply before sending his £70. As I told him, "buying" a star in that way was about as useful as paying £70 to buy a grain of sand on Bognor beach. But others have not been warned, and one registry claimed in late 1996 that it had "named" 750,000 stars. At £50 a time, this means that the perpetrators have coined a great deal of money.

Even political leaders were taken in. In 1996 a mad gunman massacred sixteen children and their teacher at a primary school in Dunblane, Scotland and the *Sun* newspaper for 16 September carried the headline:

"Major buys star for Angels of Dunblane.
PM's personal Tribute"

The star concerned was, apparently, a very dim one with the official catalogue number of 53165535, somewhere in the constellation of Cygnus. Of course, John Major had no idea that he was the victim of a confidence trick, though he was soon enlightened (by me, among other people). Unfortunately the whole fiasco gave the "registry" the sort of publicity that they wanted. It was, incidentally, revealed that someone else had "bought" the same star months before, and named it after a pop singer.

The same unsavoury manoeuvre was found in September 1997, when Diana, Princess of Wales, was killed in a road accident in Paris together with her friend, Dodi al-Fayed. The London *Times* of September 3 reported that a London woman had applied to name a star "Diana — the People's Princess", and another writer in North England applied to name another star "Dodi and Diana — Eternally Loved". If these applications were accepted (and I have no doubt that they were), the organisers would make another £100 at least.

The problem is that so far as can be ascertained, this sort of thing is not actually illegal; we enter a sort of grey area, and nobody seems to be certain where the legal line can be drawn. But one thing is certain; these unscrupulous people are coining money, and taking it from those who can least afford to lose it. Should you ever receive an advertisement from one of these bogus agencies, be on your guard.

29

Travel to the Stars?

The Space Age began very suddenly. On 4 October 1957, the Russians launched their first man-made moon or artificial satellite, Sputnik 1, and showed that travel beyond the Earth was possible. Four years later Yuri Gagarin became the first man to travel beyond the atmosphere. Since then there have been space stations, trips to the Moon and rockets to all the planets in the Sun's family apart from Pluto (which, in any case, has no right to be ranked as a proper planet). The first expeditions to Mars are being planned, and it will be strange if they do not take place within the next few decades. But can we go further — to the stars?

Let us begin by seeing what is possible and what is not; everyone will have their own ideas, and I can do no more than present my own. Reaching stars would not be very wise even if it were practicable. Our Sun, which is very mild by stellar standards, has a surface temperature of almost 11,000°F (6100°C) and near its core, where its energy is being generated, the temperature rises to the unbelievable value of 27,000,000°F (15,000,000°C), and perhaps rather more. But there is no reason why other stars should not be attended by Earth-like planets. Our Sun is one of 100,000 million stars in our Galaxy, and is perfectly normal, so that it would surely be absurd to suggest that it could be unique in being the centre of a planetary system. If we could locate another Earth, moving round a sun-like star, we might well expect to find life forms similar to ours. Of course there is no proof, but it does seem very plausible.

The two stars nearest to us which are at all like the Sun, and are therefore promising candidates as planetary centres, are Tau Ceti and Epsilon Eridani. Their distances are of the order of 11 light-years, or rather less than 70 million million miles (110 million million km). Distances of this sort obviously pose great problems, and we have to admit that sending a twentieth-century type rocket there is absolutely out of the question. It would take far too long, and there is no point in discussing it further. We must think of something better.

Modern rockets travel at speeds of several miles per second. No doubt future rockets will do better, and using ion propulsion they could accelerate to tremendous rates, but there is a definite limit: the speed of light, which is 186,000 miles (300,000 km) per second. The limit is proved by Einstein's theory of relativity, which has survived every test so far made. Relativity states that the faster you go, the more you "weigh" (to be more accurate, the greater your mass), and moreover your own time-scale slows down. If you could consider travelling at the actual speed of light, your mass would become infinite and time would come to a halt as far as you were concerned, which is another way of saying that it can't be done. These effects do not start to become evident below speeds of at least 160,000 miles (257,000 km) per second, but even this is completely beyond our technology and is likely to remain so for many centuries yet.

Science fiction writers have devised numerous ways round this barrier. One is the space-ark, in which the original travellers die early in the journey, and only their remote descendants survive to make planetfall in some distant Solar System. Frankly, this is far-fetched in the extreme, and it is hard to believe that it will ever be attempted.

Then what about space-warps, time-warps and other methods used by intrepid voyagers such as Dr Who and the crew of the starship *Enterprise*? There may be possibilities here, but at the moment we have no idea of what they may be, and we have no basis upon which we can build, so that speculation is both endless and rather pointless. Neither does it seem very likely that a spaceship could enter a convenient black hole (the region round an old, collapsed star from which not even light can escape) and reappear blithely in some other part of the universe — or even in a different universe altogether.

Next, what about thought-travel and teleportation? Here we are back to sheer science fiction, and whatever one thinks about such theories, it seems rather difficult to put them into practice.

In short, all methods about which we can talk sensibly at the moment fail us so far as interstellar travel is concerned. The stars are simply too far away. And yet the situation may not be quite so clear-cut as might be thought and there are two points which, in my view at least, may be highly relevant.

The first is that if there are other intelligent beings in the Galaxy — and why not? — they could well be in advance of ourselves. The Sun is around 5000 million years old, but civilisation on Earth, in the accepted sense of the term, dates back only a few thousand years at most. In that time, mankind has progressed from cave dwelling to reaching the Moon. The Sun will not change much for several thousands of millions of years yet, and so we have plenty of time. If we manage to use our resources properly, and resist the temptation to blow each other up, there is no knowing what we may be able to achieve. Other older, wiser races may well have developed the art of interstellar travel, and if so they could come here. There is no evidence that they have ever done so and we can dismiss all the various UFO stories out of hand, but visitations are not impossible, and if we were contacted by an alien race we might learn a great deal.

Most aliens in science fiction are hostile; the trend was set by H.G. Wells in his *War of the Worlds*, and has continued ever since. To me, this seems utterly illogical. If another race has learned enough to cross interstellar space, it will certainly have abandoned futile pursuits such as warfare. I am reminded of the words of Percival Lowell in his book *Mars and its Canals*, written in 1906: "War is the survival among us from savage times, and affects now chiefly the boyish and unthinking element of the nation. The wisest realise that there are better ways for practising heroism, and other and more certain ways of insuring the survival of the fittest. It is something a people outgrow." Any alien race arriving on Earth would come in peace, not to conquer us. Incidentally, it has been suggested that we ought not to allow spacecraft to escape from the Solar System and let other races know of our existence, just in case we attract unwelcome attention. This is equally absurd. In any case, we could not remain unnoticed even if we wanted to do so; any radio astronomer within a hundred light years of us could pick up our broadcasts. (Even at this moment a radio "ham" living on a planet orbiting Beta Cassiopeiae, just over 40 light-years away, could be tuning in to one of my very early *Sky at Night* programmes!)

To return to my main theme: science fiction has a habit of turning into science fact. It is less than a century since Orville Wright made his first short hop in his primitive aeroplane; less than seventy years later Neil Armstrong went to the Moon. Indeed, the two could have met, because their lives overlapped (Orville Wright lived until 1948; I met him). And what would King Canute have said if told that within a thousand years after his death it would be possible to sit at home, push a knob, look at a screen and see men walking about on the lunar surface? He would not have taken it seriously — but we may be no further away from interstellar travel than King Canute was from television.

The crux of the matter is that if we are ever going to travel beyond the Solar System, we must wait for some fundamental breakthrough. At the moment we have absolutely no idea what it may be, and neither can we tell when it will happen. It may come this week, next year, in a hundred years time, or in a thousand years or a million years — or never. But on no account must we rule it out.

Look at the stars on the next clear night. Each one is a sun, and many of them may have inhabited worlds moving round them; some astronomer in a far away system may be looking at our Sun and wondering whether it, too, is a planetary centre. Merely because our current technology cannot show us the way to the stars, we must not be too guarded. Sooner or later we may achieve it and send our messengers to the very ends of the universe.

30

Ghost Moons

Most of the planets in the Solar System are attended by satellites. To date the Earth is known to have one, Mars two, Jupiter sixteen, Saturn eighteen, Uranus fifteen, Neptune eight and Pluto one; this gives a total of sixty-one. Only Mercury and Venus are unaccompanied. Theoretically I suppose we might tack on Dactyl, the midget satellite of the asteroid Ida, and the as yet unnamed satellite of another asteroid, Dionysius, but for the moment let us keep to the main planets.

There have also been reports of extra satellites, which have never been confirmed and which must be regarded as ghosts. Their stories are rather interesting. (There has also, of course, been one ghost planet — Vulcan — but that is another story.)

Let us begin with our own Earth. Obviously there could be no large satellite other than the Moon, because even if it were small by satellite standards it would still be a bright naked-eye object; we are considering something very insignificant indeed. And in 1846 Fréderic Petit, Director of the Toulouse Observatory in France, announced that a second satellite had been discovered. It had, said Petit, been observed by three observers — Lebon and Dassier at Toulouse, and Larivère at Artenac — on the evening of 21 March 1846. Petit calculated an elliptical orbit, with a period of 2 hours 44 minutes 59 seconds, and a distance ranging between 2219 miles (3571 km) down to a mere 7.8 miles (12.6 km). This in itself ought to have been enough to discredit the whole idea, because any object moving at less than ten miles above ground level would be affected by air resistance and would not last for long before spiralling down to destruction, as countless artificial satellites have done.

Not to be put off, Petit made some new calculations and, in 1861, announced that he had tracked down a small satellite which was affecting the movements of the real Moon. This too was so wildly improbable that nothing more would have been heard of it but for one man: Jules Verne.

Verne was arguably the greatest of all science fiction novelists and his books are worth reading even today, particularly in the original French (inevitably they lose something in translation). One of Verne's major works was *From the Earth to the Moon*, published in 1865, followed six years later by the sequel, *Round the Moon*. His three intrepid space travellers, Barbicane, Captain Nicholl and Michel Ardan, are fired Moonward in a projectile sent out from the barrel of a huge gun. Let it be said at once that though Verne was not himself a qualified scientist, he believed in keeping to the facts as much as possible. However, evidently he did not realise that the shock of starting off at escape velocity would at once turn the luckless occupants of the projectile into fine jelly, quite apart from the fact that friction against the atmosphere would have turned them into a fireball. He did get his departure velocity right, though he was wrong about the effects of weightlessness. In his story, the travellers lost all sensation of weight only when they reached the "neutral point" where the gravitational pull of the Earth exactly balanced that of the Moon. In fact, the travellers would have been weightless from the moment of departure, because they would have been in free fall.

What about a return journey? Simple: there was not to be a return. The adventurers would, they hoped, reach the Moon, but with no chance of coming back.

This would have made for a rather dismal story, so Verne hit upon an ingenious solution: use the pull of the second satellite. During the voyage they were almost hit by an enormous object which flashed past them. I quote from the novel:

"It is," said Barbicane, "a simple meteorite, but an enormous one, retained as a satellite by the attraction of the Earth."

"Is that possible?" exclaimed Michel Ardan: "the Earth has two moons?"

"Yes, my friend, it has two moons, although it is usually believed to have only one. But this second moon is so small and its velocity is so great that the inhabitants of Earth cannot see it. It was by noticing perturbations that a French astronomer, Monsieur Petit, could determine the existence of this second moon and calculated its orbit. According to him, a complete revolution around the Earth takes three hours and twenty minutes."

"Do all astronomers accept the existence of this satellite?" asked Nicholl.

"No," replied Barbicane, "but if, like us, they had met it, they could no longer doubt it. But this gives us a means of determining our position in space. Its distance is known and we were, therefore, 7480 km [4650 miles] above the surface of the globe when we met it."

In fact, a natural satellite orbiting at such a distance would have a period of around 4 hours 48 minutes, but the point was that in Verne's story the second satellite perturbed the projectile and sent it back to a landing on Earth. If you have never read the book, I recommend you to do so.

Next in the field seems to have been a Dr G. Waltemath, about whom not a great deal is known. In 1898 he claimed to have found not one new satellite of the Earth, but *several*. One of these was said to be 640,000 miles (1,030,000 km) from the surface, and 435 miles (700 km) in diameter, and though usually very faint "sometimes," wrote Waltemath, "it shines at night like a sun", and he cited an observation made by Lieutenant Greely from Greenland, 24 October 1881, which was, he claimed, a sighting of the satellite, though Greely had believed it to be the Sun (!). Waltemath also predicted a transit of his satellite across the Sun on 4 February 1898, and there were reports that twelve people in Griefswald (in fact, members of the local Post Office staff) did see a dark body crossing the solar disc. Unfortunately two experienced astronomers, W. Winkler in Jena and the Baron Ivo von Benko at Pola in Austria, were watching the Sun at that particular moment, and saw nothing at all. Still undaunted, Waltemath continued to issue predictions and ask professional observatories to confirm them.

Eventually the professionals lost patience. To quote H.H. Turner of the Oxford University Observatory:

"Waltemath might be tolerated as comparatively harmless were it not that one has to answer so many times the genial query, 'Oh! By the way, what about those new moons?' and even to write letters about them. The time may come when people with new theories will have to be made to deposit a substantial sum, to be forfeited in case the bubble bursts, and those who want to have questions answered must pay a consultation fee of two guineas. It would probably make them value the replies a little more; and one could buy some nice piece of apparatus with the money obtained. But I fear that this will not be arranged before the next century."*

The last word on Waltemath came from an astrologer named Sepharial in 1918. He accepted the reality of the second satellite, and gave it a name: Lilith. He believed it to have about the same mass as the Moon, and to be invisible for most of the time, so that it could be glimpsed only when near opposition or in passing in transit across the Sun. It did not seem to occur to him that a body with lunar mass would make its presence very much felt. Yet even today one occasionally finds the name of Lilith in astrological horoscopes ...

On a very different tack, Clyde Tombaugh, the discoverer of Pluto, once carried out a systematic photographic search for minor satellites, with a total lack of success. However, he did establish that nothing of the Lilith type could possibly exist.

There was another flurry of interest in 1954, when there were reports of the discovery of two satellites, one orbiting at a distance of 435 miles (700 km) and the other at 620 miles (998 km). I have never been able to track down the origins of these reports, but they faded out after the launch of the first artificial satellite, Sputnik 1, on 4 October 1957.

Between 1966 and 1969 John Bargby, in America, claimed to have seen at least ten small satellites telescopically. He computed orbits for them and suggested that they might be fragments of a

---

* I have every sympathy with Professor Turner, because never a week passes without my being sent new cosmological theories for analysis. I now have the answer. I ask the senders to submit full, rigorous mathematical details. As none of the theorists ever know any mathematics, this is quite safe, and nothing more is heard from them.

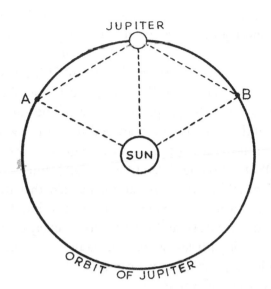

JUPITER

A

B

SUN

ORBIT OF JUPITER

The Trojan aster-
oids move in the
same orbit as
Jupiter, either 45
degrees ahead (A)
or behind (B).
These are stable or
Lagrangian points.
It is claimed that the
Kordylewski clouds
occupy analogous
positions in the
orbit of our Moon,
though the exist-
ence of the clouds
remains to be
proved.

larger body which broke up in late 1955. He based his ideas on alleged perturbations of the motions of artificial satellites, but the data he used were no more than approximate and could not possibly be of any real help. Moreover, any such satellites ought to have been naked eye objects. They have never been confirmed, and must certainly join Lilith in the realm of myth.

There is always a chance that a very small natural satellite might exist, and there is the case of the meteoroid of 10 August 1972, which entered the Earth's atmosphere, descended to a mere 36 miles (60 km) above ground level, and then moved away again. It attained magnitude -15, and may have been well over 200 feet (61 m) in diameter. Presumably it re-entered a solar orbit and is still moving round the Sun, but its existence shows that there could well be tiny bodies of the same sort in orbit round the Earth. However, the only positive results relate to what are called the Kordylewski clouds.

If a body moved round the Earth in the same orbit as the Moon, keeping either 60 degrees ahead of the Moon or 60 degrees behind it, it would be stable, and there are instances of the same sort of thing in the Solar System. The Trojan asteroids move in the same path as Jupiter; one asteroid is known to be a

Martian "Trojan", and there are also small satellites sharing in the orbits of Tethys and Dione, which are relatively large satellites of Saturn.

In 1956 K. Kordylewski and his colleague M.Winiarski suggested that there might be "lunar Trojans", and in October 1956 Kordylewski reported a bright patch in one of the stable positions. In March and April 1961 he managed to photograph two cloudy patches in the Trojan sites; they were again reported in 1975 from the OSO (Orbiting Solar Observatory) and were photographed by Winiarski in 1990. It cannot be said that the evidence is conclusive, because the patches are so very dim, but the Kordylewski clouds may exist, and they are more likely to be aggregations of very small particles rather than definite satellites. At the moment this is about as far as we can go.

Next, Neith, the satellite of Venus.

Logically there seems no reason why Venus should not have a satellite, and the first report came in 1686. It was made by G.D. Cassini, an Italian astronomer who was called to France to become Director of the new Paris Observatory. Nobody could doubt Cassini's skill. He was responsible for the discoveries of four of Saturn's satellites (Iapetus, Rhea, Dione and Tethys) as well as the main gap in the ring system, still known as the Cassini Division. His report reads as follows:

"1686 August 18, at 4.15 in the morning. Looking at Venus with a telescope of 34 feet [10.4 m] focal length, I saw at a distance of 3/5 of her diameter, eastward, a luminous appearance, of a shape not well defined, that seemed to have the same phase with Venus, which was then gibbous on the western side. The diameter of this object was nearly one-quarter that of Venus. I observed it attentively for 15 minutes and, having left off looking at it for four or five minutes, I saw it no more; but daylight was by then well advanced. I had seen a like phenomenon, which resembled the phase of Venus, on 1672 January 25, from 6.52 in the morning to 7.02 am when the brightness of the twilight caused it to disappear. Venus was then horned, and this object, which was of a diameter almost one quarter that of Venus, was of the same shape. It was distant from the southern horn of

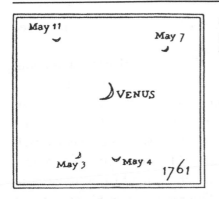

Positions of the satellite of Venus (Montaigne, 1761). In fact no satellite exists.

Venus a diameter of Venus on the western side. In these two observations, I was in doubt whether or not it was a satellite of Venus, of such a consistence as not to be very well fitted to reflect the light of the Sun, and which in magnitude bore nearly the same proportion to Venus as the Moon does to the Earth, being at the same distance from the Sun and Earth as was Venus, the phases of which it resembled."

It was recalled that F. Fontana had seen something similar as far back as 15 November 1645. Then, near sunrise on 23 October 1740, the satellite was recorded by James Short, the well-known telescope maker:

"Directing a reflecting telescope, of 16.5 inches focus (with an apparatus to follow the diurnal motion) toward Venus, I perceived a small star pretty nigh upon her, upon which I took another telescope of the same focal distance, which magnified about 50 to 60 times, and which was fitted with a micrometer, in order to measure the distance from Venus; and found its distance to be about 10°2'.0. Finding Venus very distinct, and consequently the air very clear, I put a magnifying power of 240 times and, to my great surprise, I found this star put on the same phase with Venus. Its diameter seemed to be about a third, or somewhat less, of the diameter of Venus; the light was not so bright or vivid, but exceeding sharp and well defined. A line, passing through the centre of Venus and it, made an angle with the equator of about 18 or 20 degrees. I saw it for the space of an hour several times that morning, but the light of the Sun increasing, I lost it about a quarter of an hour after eight. I have looked for it

143

every clear morning since, but never had the good fortune to see it again."

Further confirmation came in 1759 when Mayer saw it about 1 to 1½ diameters away from Venus. In 1761 Venus passed in transit across the face of the Sun and a German astronomer, A. Scheuten, recorded a small black speck following the planet across the solar disk, remaining visible even when Venus itself had passed off the Sun. Also in 1761, Montaigne of Limoges made a series of observations which seemed to be most convincing; he claimed that the satellite remained visible even when Venus itself was outside the field of view, and that he had taken every possible precaution against optical illusion. In a memoir read to the French Académie des Sciences, M. Baudouin announced:

"The year 1761 will be celebrated in astronomy, in consequence of the discovery that was made on 3 May of a satellite circling round Venus. We owe it to M. Montaigne, member of the Society of Limoges ... We learn that the new star has a diameter one-quarter that of Venus, its distance from Venus almost as far as the Moon from the Earth, and has a period of 9 days 7 hours."

On 3 and 4 March 1764 the Danish astronomer Roedkier, from Copenhagen, saw the satellite again, so did Christian Horrebow on 10 and 11 March, and Montbaron, from Auxerre, on 28 and 29 March. And henceforth the satellite disappears from the observation books. Schröter could not find it; neither could the great William Herschel. Yet its existence was regarded as definite, and in 1773 the German astronomer, J. Lambert, calculated an orbit; he gave the distance from Venus as 259,000 miles (417,000 km), and the period 11 days 5 hours. Frederick the Great of Prussia proposed to name the satellite "D'Alembert", after his old friend Jean D'Alembert, but the prudent mathematician declined the honour with thanks.

Doubts were already creeping in. Venus is so brilliant that it is prone to produce telescopic "ghosts", due to slight defects in the optics, and this was the view expressed in 1766 by Maximilian Hell, Director of the Vienna Observatory. Yet old myths die hard, and in the nineteenth century Admiral W.H. Smyth, author of the famous *Cycle of Celestial Objects*, suggested that "the satellite is perhaps extremely minute, while some parts of its body may be less capable of reflecting light than others". This idea was developed in 1875 by F. Schorr, who went so far as to write a small book about it. Schorr revised Lambert's period of 12d 4h 6m, and argued that the many failures to see the satellite were due to the fact that it varied in brightness, and was normally too faint to be visible.

This seemed improbable, and in any case it was necessary to define just what was meant by a "satellite". Whereas Cassini, Montaigne and others had described it as being a quarter the size of Venus itself, Roedkier and Horrebow had seen it as a starlike point.

M. Houzeau, Director of the Brussels Observatory, had a different idea in 1884; he believed that the body was not an actual satellite, but an independent body orbiting the Sun in a period of 283 days, so that it made periodical close approaches to Venus. It was he who proposed a name for it: Neith. But the whole matter was effectively cleared up in 1887 by Paul Stroobant, of Brussels. He analysed all the observations and explained them either as telescopic ghosts (as Father Hell had maintained) or as known stars; Roedkier's observations corresponded to the stars Chi Orionis, M. Tauri, 71 Orionis and Nu Geminorum, while Horrebow's fitted in well with the fifth magnitude Theta Librae. Lambert's orbit was clearly erroneous; it would require the mass of Venus to be ten times greater than it really is.

Only one more report was ever made — on 13 August 1892 by Edward Emerson Barnard, who was using the 36-inch Lick refractor and saw a seventh-magnitude starlike orbit in the same field as Venus. Barnard measured its position and found that it did not agree with any star. Certainly he would not have been misled by a telescopic ghost, and the observation remains unexplained, but it is possible that Barnard saw a nova or "new star" which, by bad luck, was not observed elsewhere.

Had a satellite existed, it would certainly have been found by now.

The satellite of Mercury had an even briefer existence. In fact its story began and ended within a period of less than a week.

The observations came from Mariner 10, the only spacecraft which has so far made contact with Mercury.

Mariner 10 was launched on 3 November 1973; it bypassed Venus on the following 4 February and then swung in to rendezvous with Mercury, making three active passes before its power failed.

The first of these was on 29 March 1974, the second on 21 September 1974 and the third on 16 March 1975. Almost half the surface of the planet was mapped, and showed it to be superficially very like that of the Moon, with mountains, craters, valleys and ridges. Contact was finally lost on 24 March 1975, though no doubt Mariner is still in solar orbit and still making regular close passes of Mercury.

On 29 March one of the on-board instruments, capable of observing in the extreme ultra violet part of the spectrum, saw bright emissions which, in the words of a member of the team, "had no right to be there". Later they were absent but then, after closest approach, they were back, and moreover the object — if object it really were — had moved away from the planet. The research team at the Jet Propulsion Laboratory made haste to work out the object's speed with respect to Mercury; it was consistent with what would be expected of a satellite. Yet the scientists were not convinced, and decided to wait before making any announcement.

By bad luck the Press learned about it, and premature reports appeared. The furore did not last for long. The object headed straight out from Mercury and was soon found to be an ordinary star, 31 Crateris.

Mars is smaller than the Earth but much larger than the Moon, so that satellites would not seem improbable. In 1643 one observer, Anton Schyrl, claimed to have seen one, but there is no doubt that what he observed was an ordinary star.

But the question of Martian attendants really came to public attention in 1727 when Jonathan Swift published *A Voyage to Laputa*, one of the famous travels by Dr Lemuel Gulliver.

Laputa was an airborne island which surely qualifies as the first fictional flying saucer. The Laputan astronomers were so

Laputan astronomers surveying Mars! An old woodcut from Gulliver's Travels.

skilful that they had discovered "two lesser stars, or satellites, which revolve about Mars, whereof the innermost is distant from the centre of the primary planet exactly three of his diameters, the outermost five; the former revolves in the space of ten hours, the latter in twenty-one and a half". The rotation period of Mars itself was already known to be just over 24½ hours, so that according to the Laputans the inner satellite had an orbital period of less than a Martian day. This would make it unique in the Solar System.

A Voyage to Laputa was published in 1726, and at that time there was no telescope in existence powerful enough to show the two dwarf satellites which really exist. Swift died in 1745. Five years later the French novelist Voltaire wrote *Micromégas*, in which he credited Mars with two moons. All sorts of curious speculations followed. Could it be that people of ancient times

— Atlanteans, perhaps? —had powerful instruments? In fact, the real answer is simple enough, and Voltaire himself pointed it out. Mars is further away from the Sun than we are, so that it cannot possibly manage with less than two moons.

Periodical searches were made. In 1747 a German amateur named Kindermann claimed to have seen a satellite. He gave the date of his observation as 10 July 1744, and worked out an orbital period of 59d 50m 6s, but if he saw anything at all it was undoubtedly a normal star. William Herschel looked unsuccessfully in 1783, and so, in 1862 and 1864, did Heinrich D'Arrest, co-discoverer of Neptune.

The two genuine satellites, Phobos and Deimos, were found in 1877 by Asaph Hall, from Washington. Phobos has a longest diameter of less than 20 miles (32 km) and Deimos less than 10. In all probability they are captured asteroids rather than bona fide satellites, and future colonists will find them of very little use in illuminating the darkness of the Martian night. Phobos does indeed move round Mars in less than a Martian day, so that an observer on the planet will see Phobos rise in the west, gallop across the sky and set in the east 4½ hours later. But here too there is no mystery as far as Swift is concerned; any satellite would have to be very close to Mars, or it would have been discovered even with the feeble telescopes of the eighteenth century.

Jupiter has one phantom moon, reported in 1975 by Charles Kowal from Palomar. It was seen several times but was then lost. It may, or may not, have been a very small satellite, and if so it will no doubt be recovered at some time in the future.

Saturn has the distinction of having the largest satellite family. Eighteen members are now known. Of these, the largest, Titan, was discovered by Christiaan Huygens as long ago as 1655. It is bright enough to be seen with good modern binoculars, and is larger than the planet Mercury. Cassini discovered Iapetus in 1671, Rhea in 1672, and Tethys and Dione in 1684. William Herschel added Mimas and Enceladus to the list in 1789. The much smaller Hyperion was the discovery of W. Bond in 1848, and in 1898 W.H. Pickering identified Phoebe, which is a long way from Saturn — over 8,000,000 miles (13,000,000 km) — and has retrograde motion, so that it is almost certainly an ex-asteroid. Phoebe was, incidentally, the first planetary satellite to be found by means of photography.

Were there others? In April 1861 Hermann Goldschmidt, dis-coverer of several asteroids, announced that he had found a ninth satellite orbiting between the paths of Titan and Hyperion. He named it Chiron*, but it was never confirmed. However, in 1905 Pickering announced the discovery of a satellite in the same region, and he named it Themis; the period was given as 20.85 days and the mean distance from Saturn as 907,000 miles (1,460,000 km) only slightly less than that of Hyperion. Since Pickering was one of the world's leading planetary observers, his report was taken very seriously indeed, and Themis was added to the official list of satellites, but it was never confirmed, and there now seems little doubt that it was simply a star.

I cannot resist adding a note here, because I was involved.

In 1966 Saturn's rings were edgewise on to us, and this was an ideal time to look for small inner satellites. Audouin Dollfus, at Paris, reported one, and it was named Janus. It was then lost, and at normal times, when the rings are better presented, it would be very hard to see. Later, after the Voyager 1 pass of Saturn in 1980, two tiny inner satellites were found, which periodically exchange orbits — a sort of cosmic musical chairs. Dollfus had observed one of these; they are now named Janus and Epimetheus.

I come into the story because in 1966, when making observa-tions with the 10-inch refractor at Armagh Observatory, I recorded an object which was certainly Janus. As I did not rec-ognise it as being new, I can claim absolutely no credit!

Uranus, the green planet discovered by William Herschel, is now known to have eighteen satellites. Only five of these were known before the spacecraft Voyager 2 passed by the planet in 1986. The two largest, Titania and Oberon, were found by

*Note that Chiron is the name of the asteroidal body numbered 2060, which spends most of its time between the orbits of Saturn and Uranus; but Chiron was not discovered (by Kowal) until 1977. Note also that the satellite of Pluto has been named Charon. It is only too easy to become confused.

Herschel himself in 1787, Ariel and Umbriel by the English amateur William Lassell in 1851, and Miranda by Gerard Kuiper in 1948. But there have been ghosts here too, and for once it seems definite that the great William Herschel was misled.

In the ordinary form of the reflecting telescope — the Newtonian — the light from the target object is caught by a paraboloid mirror and sent back up the open tube on to a smaller, flat mirror inclined at an angle of 45 degrees. The flat mirror sends the light-rays into the side of the tube, where an image is formed and is magnified by an eyepiece in the usual way. The drawback, of course, is that every reflection loses a little light, and Herschel decided to try a new system.

Using his favourite reflector, which had a focal length of 20 feet (6.1 m), he removed the flat mirror and tilted the main speculum so that the image of the object under scrutiny could be seen directly through the eyepiece from a cage or platform fixed to the upper end of the tube. This is always known as the Hersdhelian or front-view system. It sounds good, but in fact it is not; quite apart from difficulties of adjustment, spurious images can often be introduced. Initially, Herschel was unaware of this. In January 1790 he tested the telescope by looking at Uranus, and saw several satellites. Eventually he claimed that he had found seven. Of these two, Titania and Oberon, are genuine; another, reported in 1802, just may have been Umbriel, but the evidence is far from conclusive, and the observation could not be checked independently, because nobody else had a telescope anything like as powerful as Herschel's. The rest can only have been stars.

In 1861 G.F. Chambers, in his classic *Handbook of Astronomy*, gave a table of the satellite family as it was then believed to be:

Satellite	Discovery	Mean distance from Uranus (miles/km)	Period (days)	Notes (added by author)
1. 1851:	Lassell	128,340 (206,542)	2.52	Ariel
2. 1802:	Herschel	178,882 (287,881)	4.14	Umbriel?
3. 1790:	Herschel	226,520 (364,546)	5.89	(non-existent)
4. 1787:	Herschel	293,422 (472,214)	8.71	Titania
5. 1794:	Herschel	342,411 (551,054)	10.98	(non-existent)
6. 1787:	Herschel	404,937 (651,679)	13.46	Oberon
7. 1790:	Herschel	785,047 (1,263,403)	38.08	(non-existent)
8. 1794:	Herschel	156,992 (252,653)	107.69	(non-existent)

150

In the 1889 edition of Chambers' book it was accepted that the extra satellites reported by Herschel do not exist. Umbriel was definitely seen by Lassell in 1851, and there has been claim that one of the inner satellites (either Ariel or Umbriel) may have been glimpsed by the Russian astronomer Struve from the Dorpat Observatory, in Estonia, in 1847, but again the evidence is very vague.

No doubt extra small satellites of the outer planets do exist, and will be discovered by future space probes, but it seems that the ghosts described in this chapter will remain — well, just ghosts.

31

Ripples of Creation

According to most astronomers, everything — space, time, matter — came into existence at one moment, between 15,000 million and 20,000 million years ago (perhaps slightly earlier). If we can agree about this, we can work out a complete sequence, ending up with you and me. But we have to start somewhere.

Just how or why the Big Bang happened is something that we simply do not know. I have always compared the situation with that of an intelligent being from, say, Alpha Centauri C who pays a brief visit to Earth and spends half an hour in Bognor Regis High Street. He will see babies, children, men and women and old men and women. If he is clever enough, he will realise that a baby turns into a child, a child into an adult and an adult into an elderly person, so that he will be able to work out the life cycle of a human being. But unless someone has told him about the facts of life, he will not know how the baby got there. And this is our present situation with respect to the universe. Our "baby" is the Big Bang.

At least we can do our best to work out what went on after the Big Bang, and here we are on much firmer ground. Initially the temperature must have been fantastically high. As the universe grew colder, it cooled; galaxies, stars and planets formed, and life appeared, certainly on Earth and no doubt elsewhere also. The universe gradually assumed its "modern" form.

In 1965 two American radio astronomers, Arno Penzias and Robert Wilson, were making observations with an antenna shaped like a horn. They were conscious of a peculiar background hiss, which they could not at first identify. They believed that it might be due to pigeon droppings in the antenna, but when the droppings were cleared the hiss remained. Finally it was found that the hiss was due to the remnant of the Big Bang radiation; it indicates a general temperature of 3 degrees above absolute zero. (Absolute zero is the coldest temperature there can possibly be: -460°F (-273°C).) The discovery of the background radiation was hailed as one of the greatest achievements of the twentieth century but, strangely, it appeared to be absolutely uniform — the same from all directions. The radiation, recorded as a hiss at millimetre and sub-millimetre wavelengths, was smooth. Yet the universe today is not smooth, and herein lay the problem. How did material which was originally spread around in so uniform a way start to clump together to produce galaxies?

Then along came COBE, the Cosmic Background Explorer, a spacecraft built in the United States for the sole purpose of studying the hiss. It carried three main instruments, two of which were cooled down to a temperature of below two degrees above absolute zero, so that their sensors could not be blinded by their own heat emission. It was launched into an orbit which carried it round the Earth at an altitude of around 580 miles (930 km). The orbit was circular and took COBE over the poles, so that it was kept in use for most of the time. By careful design, it was hoped to produce temperature maps of the sky accurate to within 0.0003 of a degree, far better than anything achieved before. And sure enough, temperature differences were found: ripples which indicated slight "clumping", rarefied wisps of material, at immense distances, described at the time as "the largest and most ancient structures in the entire universe".

According to what is termed the inflationary theory, there was a very rapid expansion of the universe immediately after the Big Bang. (Remember that the Big Bang created space as well as time and matter so that it happened "everywhere".) If the spreading out of matter had been as uniform as it had been thought to be before the COBE results, there could have been no conceivable way in which galaxies could have formed, so that it is fair to say that the COBE revelations came as a great relief to astronomical theorists.

Remember, too, that we are looking far back in time. The phenomena recorded by COBE date back to little more than 300,000 years after the Big Bang itself.

How did astronomers react to the news? In general, they were ecstatic. I quote:

"It is the discovery of the century, if not of all time." *Professor Stephen Hawking.*

"This is absolutely tremendous. It's a landmark in cosmology... this is a tremendous triumph of the human intellect." *Professor Michael Rowan-Robinson.*

"This has been greeted with much joy by the cosmology community, because it's very much in line with what has been predicted." *Professor Ken Pounds.*

"This is a wonderful discovery for us. It's a confirmation of a picture of the hot Big Bang universe to which we have subscribed for some time." *Dr Jasper Wall, Director of the Royal Greenwich Observatory.*

And from the *Astronomer Royal, Professor Sir Martin Rees*: "People have been looking for these fluctuations for twenty years. Their discovery strengthens our view that we're looking along the right route in trying to understand how galaxies are formed."

But there must always be someone to pour cold water on the proceedings, and who better than a senior official of the Church? The *Rt. Rev. Bill Westwood, Bishop of Peterborough*, stepped nobly into the breach. "This doesn't make a great deal of difference to me. It certainly doesn't make any difference to God."

Well, he may be right.

32

"Des. Res." on Mars?

Would you like to go for an afternoon's fishing in a lunar sea? Or build a home on a Martian plain, far from the madding crowd, with a golden landscape stretching in all directions and a majestic volcano in the background? According to some enterprising agencies, nothing is easier. All you have to do is pay a deposit, file a claim, and all will be well.

Or — will it?

There are problems to be faced. You will have to be ready for a journey of a quarter of a million miles for the Moon, and a great deal longer for Mars, and since there will be no local corner shops you will have to take all your own provisions as well as your own air. For the Moon you will also have to take a good supply of water though, admittedly, there seems to be plenty of ice on Mars which can be melted. Clothing is another matter for concern, since the night temperatures plummet to well below -100 degrees, and ordinary winter woollies will not suffice.

However, organisations such as the Lunar Embassy, based in California (where else could it possibly be?) are quick to gloss over these little difficulties and point out that you will be blissfully undisturbed; neither will there be any trouble with planning authorities. Up to the present time it is estimated that the Embassy and like organisations have sold over 100,000 square miles (259,000 sq km) of the Moon at around £3 per acre, and 15,000 square miles (39,000 sq km) of Mars at £19.50 a plot. For his modest plot, the proud purchaser receives a deed of ownership, a map, and copy of the Martian Bill of Rights. The latter document is issued by the Martian Consulate, though it is true that the authenticity of this Consulate seems a little dubious; its President and Chairman is a 49-year-old ventriloquist who has previously doubled as a car salesman. Nothing has been said about passports, and customs regulations will not need to be

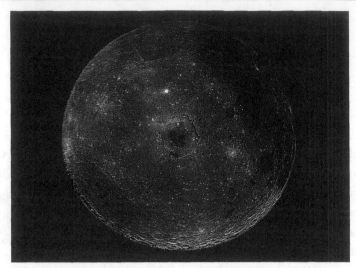

The Moon photographed from Galileo on 9th December 1990 at a range of 350,000 miles (560,000 km). In the centre is the Mare Orientale, 620 miles (998,000 km) across, which the author had the honour of discovering many years before the Space Age! The "near side" to the moon is to the right, with the vast Oceanus Procellarum; the far side is to the left. Lower left is the 1200-mile (1930-km) South Pole Aitken Basin.

rigidly applied, since there is little of financial value to be imported from the Moon and nothing more than a few bags of anorthosite from Mars.

In fact, the question of ownership may well have to be tackled eventually if, as may be the case, we establish the first Lunar Bases before the year 2020 and despatch the first Martian expeditions during the first half of the coming century. The example of Antarctica is encouraging, since teams from the various nations seem to be working there in harmony, but when we come to the Moon and Mars there will be endless regulations and red tape. As is well known, there is no complex problem which United Nations bureaucrats cannot make worse. We must hope for the best.

One point is worth making here. Bogus agencies which claim to be able to name stars are tricking gullible people — often bereaved people — out of large sums of money, but it is hard to believe that anyone can really be taken in by the prospect of buying land on another world. Frankly, it would be fascinating to go angling on the Moon, but I fear that there are no trout in the Sea of Tranquillity!

33

The Edge of the Moon

Before the onset of the Space Age, the fact that the Moon completes one orbit in exactly the same time that it takes to complete one revolution on its axis — 27.3 Earth days — was a source of intense annoyance to lunar observers such as myself. All in all we could see 59 per cent of the surface at one time or another, though of course never more than 50 per cent at any one moment. The Moon "rocks" slowly to and fro, because although its rate of axial spin is constant, its orbital speed is not. Its velocity in orbit is variable; the Moon moves fastest when closest to us. There are other less important librations, but the fact remained that almost half the lunar surface was completely inaccessible.

All sorts of theories were proposed. One was due in the mid-nineteenth century to an eminent Danish astronomer, Andreas Hansen, who believed that the Moon was not uniform in density, but has one hemisphere rather heavier than the other. This, said Hansen, would put the centre of mass 33 miles (53 km) away from the centre of the globe, and all the air and water would be pulled round on to the far side, which might even be inhabited. Few people took this seriously, and there was even less support much later for a claim by George Adamski, the first celebrated flying saucer writer, who announced that he had been taken on a round trip by astronauts from Venus, and had seen little furry creatures running about on the far side of the Moon. It was difficult to improve upon the feelings expressed in a verse allegedly written by a housemaid in the service of a well-known poet:

> O Moon, lovely Moon with the beautiful face,
> Careering throughout the bound'ries of space,
> Whenever I see you, I think in my mind
> Shall I ever, O ever, behold thy behind?

It is notable that all the large seas or maria on the Earth-turned hemisphere of the Moon are confined to the visible face. They do not spread over the edge or limb of the disc on to the far side. This led to a suggestion that the invisible hemisphere might be almost, or quite devoid of the large grey lava plains, but there was no reason to doubt that there would be plenty of craters as well as mountains, ridges and valleys. Some craters on the near side, notably Tycho in the southern uplands and Copernicus in the Oceanus Procellarum, are the centres of systems of bright streaks or rays, and efforts were made to trace rays which emanated from an origin on the far side; this might lead to at least an approximate position for a ray crater.

The first efforts in this direction were made by an American, N.S. Shaler, around 1908. He plotted six possible centres, all on the far side and therefore permanently invisible from Earth. Unfortunately he mislaid his notebooks and when he returned to the problem, years later, he found that his eyesight was no longer keen enough to make the delicate observations required. Members of the Lunar Section of the British Astronomical Association took up the problem after the end of the war; I was involved in the programme and, by then, I was spending most of my observing time in trying to plot the "edge" areas. The limb regions are very foreshortened and it is difficult to tell the difference between a crater-wall and a ridge. Maps were frankly inaccurate, even the best of them. My main references were the charts of Neison, Goodacre and Elger, but in the libration regions there were marked disagreements and many important features were not shown at all.

Obviously I could do little observing during the war because, between 1940 and 1945, I was flying with RAF Bomber Command and could get at my telescope only when on leave, which was seldom. Moreover I had nothing more powerful than a 3-inch refractor. (I still have it, and I still use it; it cost me the princely sum of £7.10s when I bought it in 1935.) My surviving notebook tells me that I did manage a reasonable view on 15 October 1940, when I used it to make the first observations of two formations right at the edge of the visible disc: a large crater now named Einstein and what turned out to be a major sea, the Mare

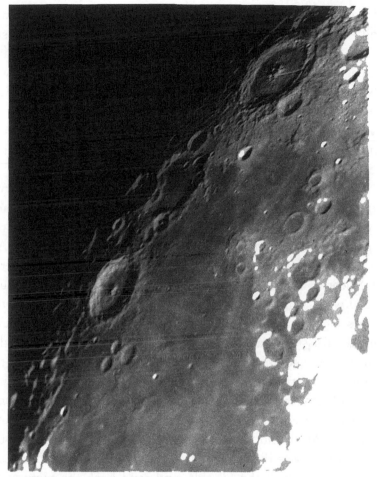

This lunar scenery shows large foreshortened craters near the Moon's limb. The three largest are Patavius (top) crossed by a cleft; the semi-ruined Vendelinus (middle) and Langrenus, with its central mountain (lower).

Orientale. These two are mainly on the far hemisphere, so that they can be seen only when libration tilts them toward the Earth, and even then only part of the Mare Orientale can be seen.

I remember that night very well. I had come home a few hours earlier, and I left a day later to return to my RAF base. It was a clear night and the Moon was brilliant. I saw what I noted as "a small sea... not on the maps — must be new", and also "a very prominent crater with high walls and a prominent central structure, which does not seem to have been reported by anyone else".

159

Unfortunately that was more or less all I could do at the time, and I was unable to return to the problem until we had disposed of the late unlamented Herr Hitler. It was only in 1947 that I managed a detailed chart of the Mare Orientale. I think I was also responsible for suggesting the name, the Eastern Sea, because it was on the eastern limb of the Moon. (Later, a decree by the International Astronomical Union reversed east and west, and the sea is now on the *western* limb, but the name has not been altered.)

At that time it was still possible to make really useful observations with a tiny telescope. I acquired a 12½-inch reflector, and continued to chart the foreshortened areas, together with the members of the Lunar Section of the British Astronomical Association. Then, in 1958, I had a letter from the USSR Academy of Sciences in Moscow. Would I please send them all my published work with the lunar libration areas, together with all the charts I had not published yet? I doubt if I really appreciated why the Russians wanted my results, but of course I complied. I sent the papers which had appeared in print, worked up all my notes into charts and sent them as well.

On 4 October 1959 the Russians launched a new probe, Lunik 3 (often referred to now as Luna 3). They had already sent Lunik 1 past the Moon, and crashed Lunik 2 on to the surface, but the new mission was different. It was meant to go right round the Moon, photograph the far side, and send back the pictures by television techniques. I was vastly intrigued. Would our preliminary results be of the right order, or would they be wildly wrong?

I had already started my *Sky at Night* series on BBC Television, and our next programme was due to be transmitted on October 24. (In those days, of course, everything was black and white, and everything was live.) We contacted Moscow, and received an assurance that the pictures would be sent over just as soon as they were available.

By 4.30 GMT on 7 October, the rocket had passed by the Moon and lay beyond it, at a distance of 40,000 miles (64,000 km) from the lunar surface. The photographic apparatus was switched on, and for the next forty minutes the pictures were taken. Two

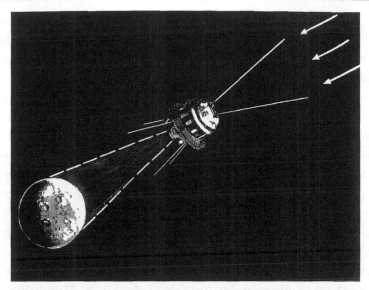

Lunik shot of the Moon photographed on 27th October 1959, showing the position of the automatic interplanetary station in space when photographing the Moon's reverse side. The arrows on the right show the direction of the Sun's rays.

cameras were used, giving photographs in different scales. After the programme had been completed, the films were automatically developed and processed, and fixed ready for transmission back to Earth.

There was bound to be a lengthy delay because Lunik 3 was still receding from the Earth. It reached its furthest point on 10 October, when it was 292,000 miles (470,000 km) away, and then started to swing in once more, reaching perigee, 29,999 miles (48,280 km) on 18 October. It was then that the pictures were transmitted. They were scanned by a miniature television camera, and were picked up by the waiting Russians. Late on 24 October they were ready for release.

In the BBC studio at Lime Grove, in Shepherd's Bush, it was quite tense. We were due on the air at a set time, and this could not be changed. The programme started. I faced the cameras — and still no pictures. Then I heard the producer's voice in my earphones: "Far-side pictures arrived. Coming on your screen in thirty seconds. Play it by ear. Good luck!"

I had no idea what we would see — and I was honest enough

to say so. Then the monitor flickered, and the first image came up. It was blurred and uncertain but, by good fortune, it did show a part of the familiar hemisphere. I recognised the well-marked Mare Crisium, seen under reverse lighting, so that I was able to give what I hoped was an intelligible commentary. Not much detail could be seen on the hitherto unknown areas, mainly because the pictures had been taken under the equivalent of full moon illumination, but clearly I had been right in maintaining that there would be fewer "seas" than on the familiar hemisphere.

When the first pictures were cleaned up and studied, it became clear that there were craters aplenty, and one fascinating structure, now named in honour of the Russian rocket pioneer Tsiolkovskii, which seemed to be a cross between a crater and a mare. There were two minor seas; the Russians also plotted a long chain of peaks and named them the Soviet Mountains, but these proved to be non-existent, the appearance was due to a bright ray, and the Soviet Mountains quietly disappeared from later maps. Ray craters existed, and several of our "centres" were found to be not too wide of the mark.

Lunik 3 could do no more. Evidently the Russians meant to re-run the pictures, but contact was lost abruptly and was never regained. It may well be that the probe is still orbiting the Earth Moon system; we will never know.

Within a few years, of course, the whole situation changed. Between 1966 and 1968 the five American Orbiters went round and round the Moon, sending back high quality pictures of the entire surface, the unknown side of the Moon was unknown no longer. But I still have the lunar globe which the Russians sent me when they had still to map parts of the averted hemisphere, and certainly I will never forget my first sight of those blurred, flickering pictures on my television monitor.

34

The Sad Case of Dr Elliott

In the year 1787, a certain Dr Elliott was brought to trial at the Old Bailey, accused of shooting a Miss Boydell. The case was reported in the *Gentleman's Magazine* for that year. No doubt this magazine has been long defunct, but fortunately the trial was also described in 1811 by David Brewster, who wrote a book on astronomy which had a wide circulation. I quote:

"The friends of the Doctor maintained that he was insane, and called several witnesses to establish this point. Among these was Dr Simmons who declared that Dr Elliott had, for some months before, shown a fondness for the most extravagant opinions and that, in particular, he had sent him a letter on the light of the celestial bodies to be communicated to the Royal Society. This letter confirmed Dr Simmons in the belief that this unhappy man was under the influence of this mental derangement. As proof of the correctness of this opinion, he directed the attention of the court to a passage of the letter in which Dr Elliott states, 'that the light of the Sun proceeds from a dense and universal aurora, which may afford ample light to the inhabitants of the surface [of the Sun] beneath, and yet be at such a distance aloft as not to annoy them. No objection,' says he, 'ariseth to that great luminary being inhabited, vegetation may obtain there, as well as with us. There may be water and dry land, hills and dales, rain and fair weather, and as the light, so the seasons, must be eternal; consequently it may easily be conceived to be by far the most blissful habitation of the whole system'."

My efforts to find out what happened at the trial met with no success, but after all was Dr Elliott alone in his views? Not so, they were shared by the greatest of the time, William Herschel, discoverer of the planet Uranus and also the man who drew up the first reasonably accurate picture of the shape of the Galaxy.

When I first read about Dr Elliott, I thought that it would be interesting to see how popular views about the Sun and its spots changed over the years, and I admit that I was surprised, though I ought not to have been. Even though solar spectroscopy began in the early nineteenth century, ideas a hundred years later were still remarkably primitive, at least insofar as sunspots were concerned.

Let us begin with William Herschel himself. This particular passage dates from 1773, but he repeated it many times and he never changed his views; he died in 1822. "The Sun appears to be nothing else than a very eminent, large and lucid planet, evidently the first, or rather primary, one of our Solar System, all the rest being truly secondary to it. Its similarity to the other globes of the Solar System, with regard to its solidity, its atmosphere and its diversified surface, lead us to suppose that it is most probably also inhabited, like the rest of the planets, by beings whose organs are adapted to the peculiar circumstances of that vast globe."

One might imagine that the Solarians would have to put up with a rather warm climate, but Herschel had the answer. The Sun's rays produce heat only when they enter an appropriate medium, such as the Earth's atmosphere....

To Herschel, then, the Sun was a dark, non-luminous body, and the heat which we receive comes from its exterior. There are two layers of cloud. The outer one is incandescent; the inner one is dark, but capable of reflecting light from its upper face, thereby acting as a screen to protect the solid, habitable body below. Sunspots are simply openings in the cloud layers. As we know, a major spot consists of a dark central portion or umbra, surrounded by lighter penumbra. According to Herschel, the umbra marks the middle of the opening, all the way down to the dark globe, while the penumbra comes from the lower cloud layer.

Even in the eighteenth century these views sounded rather extreme, and it is on record that Nevil Maskelyne, the Astronomer Royal, used to edit Herschel's papers and remove most references to the inhabitants of the Sun*. But certainly the nature of sunspots remained unknown and there were many theories about them.

One well-known astronomical writer of the time was Thomas Dick, and in his book *The Solar System*, published in 1799, we read:

"[Conclusions are] that the central part of the spots, i.e. beneath the level of the Sun's surface, or, in other words, that the spots are excavations in the body of this luminary, and that the umbra, or shade, which surrounds it is the shelving sides of this excavation in the luminous matter." Yet in the same year there appeared a book by Margaret Bryan *Lectures in Astronomy*, which gave a rather different explanation: "Less bright spots are perceived on the surface, the shape of which is variable, as well as the situation and number of them. These appearances are at first obscure, and then become brighter by degrees, so that at last they exceed the other parts of the Sun in brilliancy. Their obscurity is supposed to be produced by the smoke of the volcanoes previous to the eruption." This might be a reference to faculae, but the idea of sunspots as the tops of erupting volcanoes was quite common at the time.

However, the Scottish astronomer Alexander Wilson had made an important discovery as long ago as 1769. When a spot is close to the Sun's limb it is bound to be foreshortened, and if it is actually circular it will appear elliptical. In general, the penumbra of the spot will be narrower in the direction toward the Sun's centre than toward the direction of the limb, and this indicates that the spots must be saucer-shaped depressions. If they were humps, then the penumbra would appear narrower toward the limb. Not all spots show this Wilson Effect, but many do, and the argument is sound enough.

Ideas did not seem to change much throughout the nineteenth

---

*It is often believed that William Herschel was Astronomer Royal. He was not; true, he was appointed official astronomer to King George III, but this is not the same thing. Nevil Maskelyne was Astronomer Royal from 1765 to 1811; on his death John Pond succeeded him.

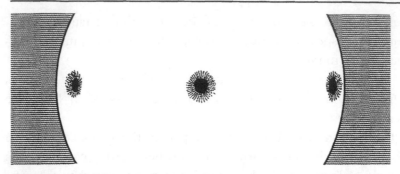

The Wilson Effect was discovered by Scottish astronomer, Alexander Wilson, in 1769.

century, though the idea of Solarians was tacitly abandoned. Admiral W.H. Smyth wrote in 1850: "By some, the dark spots are supposed to be scoria floating on the abyss of combustion, and the faculae to be volcanic eruptions from the fused mass. That view, however, which best explains these appearances, and therefore obtains credence among astronomers, regards the Sun as a black solid nucleus surrounded by two atmospheres, the one obscure, the other luminous." In 1863 William Leitch wrote a book *God's Glory in the Heavens*, in which he outlines a novel theory: "A Chinese ivory ball, composed of carved concentric shells, represents very well the structure of the Sun and the nature of the spots. In looking down the large holes in the ivory ball, we see the successive edges of the concentric shells and, in like manner, do we see the successive edges of the concentric strata of the Sun."

William Herschel's son John (who became Sir John) was an eminent astronomer in his own right, and was the first to carry out a systematic survey of the far-southern stars. During the 1830s he spent several years at the Cape of Good Hope and it is fair to say that he is the real founder of southern hemisphere astronomy. He was also a popular writer, and his book *Outlines of Astronomy* remained the standard text for many decades. No doubt he was to some extent influenced by his father's views. In the 1869 edition of *Outlines* we find: "But what *are* the spots? Many fanciful notions have been broached on this subject, but only one seems to have any degree of physical probability, viz., that they are the dark, or at least comparatively dark, solid body of the Sun itself laid bare by those immense fluctuations in the luminous region of its atmosphere, to which it appears to be subject." The strange thing about all this is that by now solar spectroscopy was well advanced, and Jules Janssen and Norman Lockyer had found out how to observe the solar prominences without waiting for a total eclipse. Yet another pioneer solar spectroscopist, Johann Zöllner, professor of astronomy in Leipzig, was maintaining that the Sun's surface is liquid, a molten mass overlaid by an atmosphere; spots are slaglike patches on the liquid surface, appearing dark because they are cooler than their surroundings.

Astronomy today is a fast-moving science; a book written even a year ago is liable to be out of date. This was not always so. In my possession I have two editions of a famous book, *The Story of the Heavens* by Sir Robert Ball, sometime Astronomer Royal for Scotland and the best-known popular author of the day. (I never knew him, because he died before I was born, but he was a close friend of my grandfather's.) One edition is dated 1893, the other 1903, and they are virtually identical. This is what he says about sunspots:

"The sunspots are variable features exhibited by that envelope of glowing clouds surrounding the Sun which we call the photosphere... We have, indeed, the best reasons for knowing that changes on the most gigantic scale, and manifested with utmost violence, are incessantly in progress in the photosphere. Thus it sometimes happens that the glowing clouds are parted asunder, and a glimpse is afforded through the opening into the solar interior. It appears almost certain that the inside of the Sun has much less power of radiating light than is possessed by the exterior, and consequently such peeps as we obtain disclose a darker surface than that of the photosphere."

The idea of a hot surface overlying a darker interior seems quaint enough now, and did not seem to be really convincing in 1903, yet Sir Robert repeated it in another book, *Starland*, published in 1914. It was probably the brilliant theoretical work by Sir Arthur Eddington, in the 1920s, which caused a change of outlook, and in 1935 the then Astronomer Royal, Sir Harold

Spencer Jones, painted a very different picture in his book *Worlds without End*:

"A sunspot is a gigantic funnel-shaped vortex in the outer regions of the Sun. Around the vortex intensely hot gas from within the Sun is whirling spirally upwards. We can compare a sunspot vortex with the hollow vortex formed by water emptying out of a bath, if we imagine the water to be made to run in the opposite direction and to stream upwards from the outlet into the bath. As the gases stream out of the funnel-shaped mouth of the vortex, the pressure which has urged them upwards is released; the emitted gases then stream more or less radially outwards from the spot along the surface, producing the radial fibrous structure seen in photographs of the penumbra of spots."

This does sound decidedly more modern, and nowadays I am quite sure that there is no astronomer who doubts that the Sun's interior is a great deal hotter than the surface. We have come a long way since the time of Sir William Herschel. But I would still like to know what finally happened to Dr Elliott!

35

Fast Lane to Pluto

It was in 1930 that my old friend Clyde Tombaugh discovered Pluto. He was making a deliberate search for a trans-Neptunian planet, using a fine refractor specially installed at the Lowell Observatory in Arizona; it had been Percival Lowell who had predicted the position of a new world. Pluto turned up very close to the position given by Lowell, but ever since then it has presented astronomers with puzzle after puzzle.

It does not seem to be a proper planet. For one thing, it is small. Its diameter is now known to be a mere 1444 miles (2324 km), and this is smaller than several satellites: our own Moon, Titan in Saturn's system, Triton in Neptune's, and all four of the Galilean satellites of Jupiter. The fact that Triton (diameter 1681 miles or 2705 km) is appreciably larger than Pluto may be significant. It was once believed that Pluto may have begun its career as a satellite of Neptune, and then broke away to go it alone. But this idea was more or less abandoned when it was found that Pluto has a companion, Charon, whose diameter is

The dome of the Lowell refractor in Arizona.

169

more than half that of Pluto and has been measured as 753 miles (1212 km). If Pluto and Charon had originally been satellites of Neptune, it is most unlikely that they would have stayed together. Pluto has a rotation period of 6 days 9 hours; this is also the time taken for Charon to complete one orbit, so that the two are locked in a most peculiar manner. To an observer on Pluto, Charon would remain motionless in the sky.

Next, the orbit. Pluto takes almost 248 years to go round the Sun, but its orbit is much more eccentric than those of the other planets. The distance from the Sun ranges between 4583 million miles (7376 million km) and only 2766 million miles (4450 million km), so that for part of its orbit is actually closer in than Neptune. This was the case between 1979 and 1999; Pluto last passed perihelion in 1989. The orbital tilt is 17 degrees, so that there is no fear of a collision with Neptune. The rotational axis is inclined at an angle of 122 degrees, more than a right angle, and the escape velocity is a feeble 0.7 of a mile (1.1 km) per second.

Obviously Pluto appears small. The apparent diameter is around 0.2 of a second of arc and, in most telescopes, Pluto appears as a starlike point. As the apparent magnitude never exceeds 13.8, you need a fair-sized telescope to see it at all. However, it is known that at the moment it has an extensive, though very thin, atmosphere, and the Hubble Space Telescope has been able to distinguish light areas and darker patches on its surface. I have even suggested a preliminary nomenclature for these features, making great play of Underworld deities such as Minos, Rhadamanthys, Aeacus and Persephone.

We now know that Triton, the senior attendant of Neptune, is a strange place. Its pole is covered with pink nitrogen snow and there are active nitrogen geysers. In 1989 Voyager 2 sent back close-range pictures, and we have an excellent knowledge of the main surface details. But Pluto is a different matter; no spacecraft has yet gone anywhere near it, and so there is a great deal about it that we do not know. Has Pluto, also, pink nitrogen snow? Is the surface covered with ice? What is the composition of the atmosphere? And is Charon similar to Pluto, or is it quite different?

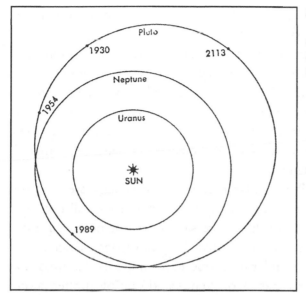

Between 1979 and 1999 Pluto's distance from the Sun was less than that of Neptune.

Even the Hubble telescope cannot answer these questions. The only way to find out the answers is to send a spacecraft. Moreover, there is need for haste. Pluto is now drawing away from the Sun, and when it moves into the further part of its orbit it will become so cold that its atmosphere will freeze out. For part of its long "year" Pluto must be airless, and we are anxious to obtain some reliable data before the atmosphere collapses.

Of course the problem is financial. Spacecraft cost money — not a great deal by national standards (and less than a nuclear submarine or two) — but with the current cutbacks in funding the prospect of a Pluto probe has seemed to be remote — until now. However, NASA has come up with a new scheme. Originally designated the Pluto Fast Flyby (PFF), the new Pluto Express is scheduled to be a two-spacecraft mission. The aims are quite clear-cut. The probes will:

1. Characterise the global geology and geomorphology of both Pluto and Charon.
2. Map Pluto's surface.
3. Determine the composition of Pluto's surface materials (and those of Charon too, if possible).
4. Determine the structure and composition of Pluto's atmosphere. Charon, with an even lower escape velocity, seems to

have no atmosphere at all, but this is something to be checked by the Express.

The overall structure of each probe will be an aluminium hexagonal "bus" with no deployable structures. The original mass will be 187 pounds (85 kg), and power will be provided by RTGs (Radioisotope Thermal Generators) similar to those used on earlier missions to the far reaches of the Solar System, where solar power cannot be used for the excellent reason that there is not enough sunlight.

Launch date? Well, 3 March 2001 has been suggested but this depends upon funds, and in any case the journey will not be at all straightforward. It will be necessary to use the now well-tested "gravity assist" technique, so that the pull of one planet can be used to give the wandering probe extra impetus. Express may have to make three passes of Venus and one of Jupiter; alone, the launching vehicle (either an American Delta or a Russian Molniya) would not be nearly powerful enough to send Express all the way to Pluto.

Of course, there will be on-board instruments of all kinds, and these will begin their main work between 6 and 9 years after launch. By then the spacecraft should be within range of its target. Attitude control should not be a problem and communications will be via a high-gain antenna 4½ feet (1.37 m) across.

What will Express tell us, assuming that all goes well? Pluto may be similar to Triton, with pink snow and nitrogen geysers. On the other hand it may be totally different, and about Charon our present ignorance is more or less complete, so that it is rather pointless to speculate. It is a pity that we have so long wait. It is not likely that Express will fly past Pluto and Charon until the year 2007 at the earliest, and probably 2012 is a more realistic estimate. I doubt whether I will still be presenting *The Sky at Night* at that time — by 2012 I will have reached the advanced age of eighty-nine — but no doubt my successor will have a great deal to say. Pluto Express seems set to be a particularly fascinating mission, and I very much hope that it does not become yet another victim of the politicians who are always so anxious to prevent our reaching out to explore other worlds.

36

Life *Can* Appear — but *Will* it?

Mars has always been of special interest to us. It is less unlike the Earth than any other planet in the Solar System, and has long been regarded as a possible abode of life. True, Percival Lowell's canal-building Martians have long since been banished to the realm of myth, but we cannot yet be sure that Mars is, and always has been, completely sterile. Certainly there must once have been liquid water there, so that in the remote past Mars was much less unwelcoming that it is today; we may be seeing the Red Planet at its very worst.

The Viking missions of the 1970s showed no definite signs of life, though it must be admitted that the results were not clear-cut; there is something decidedly peculiar about Martian chemistry. In 1997 we had Pathfinder and Global Surveyor; neither of these could be expected to detect life, but they paved the way for the probes of the next few years. Within a decade it ought to be possible to send a probe to Mars, collect material and bring it home for analysis in our laboratories. Then, with any luck at all, we ought to find out whether any Martian organisms have ever existed. Just in case they have existed (or still do), the samples will be rigorously quarantined until they have been shown to be harmless, and I suspect that the first analyses will be carried out in spacecraft.

If any Martian life is found, it will be very lowly; there is no chance of anything so advanced as a blade of grass. So why is its detection — or non-detection — so important? One special reason occurs to me, and I have not seen it well emphasised before. My question is this. If life is possible elsewhere, can we assume that it will occur? This is where Mars can give us the key.

The Sun is one of 100,000 million stars in our Galaxy alone. It is a very common sort of star, and there are many of just the same type. Moreover, we know of at least 1000 million galaxies, so that the total number of stars in the universe is unbelievably large. Can we assume that our own, utterly undistinguished Sun, is the only one to be the centre of a planetary system? This is surely absurd, and in any case we do now have strong indirect evidence of the existence of extra-solar planets. Many of these must be similar to Earth and therefore could support life. What we do not know, as yet, is whether life will automatically appear wherever conditions are suitable for it.

And herein lies the importance of Mars. If Mars has ever supported life, no matter how lowly, it will go a long way toward proving that life merely awaits a suitable environment before gaining a foothold, and will evolve as far as it can before the conditions become too hostile. If life can be shown to have appeared independently on two planets in the system, Earth and Mars, then there must be a strong argument that life is widespread elsewhere too.

If no signs of life are found, the situation will be less clearcut, and we cannot jump to conclusions by saying that life elsewhere is likely to be uncommon. We will need further evidence; the jury will still be out.

I believe that this whole problem is more important than most people realise, and, at the moment, I have a completely open mind. Frankly, I hope that the probes of the next few years will show that Mars does, or once did, support primitive life, but in my view the chances are about fifty-fifty. Time will tell.

37

Thatcher's Comet

Most comets are named after their discoverers; thus the spectacular visitor of 1997, discovered independently by Alan Hale and Thomas Bopp, was known as Hale-Bopp. A few honour the mathematicians who first computed the orbits; the most celebrated example of this is, of course, Halley's Comet. Others are the comets of Encke (J.F. Encke, sometime Director of the Berlin Observatory) and Crommelin (A.C.D. Crommelin, late of Greenwich). Politicians do not enter into it at all. I say this because there is one comet, Thatcher's, that was once linked in the popular Press with the British Prime Minister. I can assure you that there is no connection whatsoever. And yet this comet is of interest, because it is associated with an annual meteor shower, the April 1 Lyrids.

An amateur astronomer, A.E. Thatcher, who lived in New York, discovered the comet on 5 April 1861. The comet was then of magnitude 7.5, that is to say too faint to be visible with the naked eye but easy in binoculars. Thatcher described it as "a tailless nebulosity, 2 arc minutes in diameter, with a central condensation". It brightened slowly during April and, by the end of the month, had come within naked eye range and developed an appreciable tail. It continued to grow brighter, and between May 9 and 10 the magnitude had risen to 2.5 and the tail was very obvious indeed. It was then less than 30,000,000 miles (48,000,000 km) from the Earth, which cosmically is not very far, but it faded as it drew away from us. It passed through perihelion on 3 June, at 85,400,000 miles (137,650,000 km) from the Sun — well inside the Earth's orbit — and then began its outward journey. It moved south in the sky and was followed for some weeks. Apparently the last sighting of it was on 7 September when the magnitude had fallen to about 10.

So far as I know, no sketch of the comet exists, and of course this was before the days of effective photography, but the observations had been accurate enough for an orbit to be worked out. The period was established as about 415 years; at its furthest from the Sun the comet must recede to a distance of 55.7 astronomical units, or approximately 5200 million miles (8369 million km). If this is correct (and we have every reason to suppose that it is) the comet was on view about the year 1446, but records of that period are not detailed enough to help in identification. Moreover, the comet was not a brilliant object and may well have been overlooked. It should be back in or about 2276, and no doubt astronomers of the time will be on the watch for it.

But why this interest in Thatcher's Comet which was, by no stretch of the imagination, "great"? The answer lies in its association with the Lyrids.

Meteors are cometary debris. As a comet moves along, it leaves a dusty "trail" behind it, and when the Earth passes through such a trail it collects a large number of particles, each around the size of a grain of sand. When a particle dashes into the upper air, moving at anything up to 45 miles (72 km) per second, so much heat is caused by friction against the atmosphere that the particle burns away, producing the streak of radiance which we term a shooting-star. What we see, of course, is not the particle itself, but the effects it produces in the upper atmosphere during its headlong plunge to destruction.

In some cases the parent comet may spread debris all round its orbit. This has happened with, for instance, Comet Swift-Tuttle, so that we pass through the trail every August and are treated to a shower of the Perseid meteors. In other cases the particles are bunched up, so that we see spectacular displays only when we pass through the thickest part of the trail, as with Comet Tempel-Tuttle, parent of the November Leonids. In general we see a good Leonid display only every 33 years, which is the time taken for the comet to complete one orbit round the Sun. But most "parent comets which have been identified are of short period; Thatcher's is in a class of its own.

The Lyrid shower usually begins on or about 16 April each

year but does not become noticeable until 19 April. It peaks on 21 April, when the ZHR or Zenith at Hourly Rate is usually about 10; that is to say, about ten Lyrids per hour would be expected to be seen by a naked eye observer under ideal conditions, with the radiant at the zenith or overhead point. (In practice these ideal conditions are never attained, so that the observed rate of any shower is always rather less than the theoretical ZHR.) Now and then the Lyrids are rich, and reports dating back to the year 1803 state that on the night of 19-20 April there was a true meteor storm with a rate of 700 per hour. This has never been repeated, but they were much more numerous than usual in 1945 and in 1982, and they are always worth monitoring.

It was in the 1860s that astronomers first definitely identified some meteor showers with parent comets. In Vienna, Professor Edmond Weiss made careful calculations, and found that the orbits of the Lyrids fitted in very well with that of Thatcher's Comet. In the same year Johann Galle (one of the co-discoverers of the planet Neptune) confirmed the association, and traced the history of the Lyrid shower back as far as 16 March, 687 BC.

Many other comet/shower relationships are known. Among them are:

Shower	Maximum	Usual ZHR	Parent comet	Period of comet years
Lyrids	21 April	10	Thatcher	415
Eta Aquarius	5 May	35	Halley	76
Ophiuchids	9 June	5	Lexell?	now 280
Perseids	12 August	75	Swift-Tuttle	133
Orionids	21 October	20	Halley	76
Draconids	10 October	var.	Giacobini-Zinner	6.5
Taurids	3 November	10	Encke	3.3
Leonids	17 November	var.	Tempel-Tuttle	33
Andromedids	20 November	v. low	Biela	6.6
Geminids	13 December	75	(Phaethon	1.4)
Ursids	23 December	5	Tuttle	13.5

No parent comet has been traced for one of the major annual showers, the January Quadrantids. Lexell's Comet used to have a period of less than 10 years; it was seen only at one return —

that of 1770 — and its orbit has now been so perturbed by Jupiter that so far as we are concerned it is hopelessly lost. The period of 280 years may or may not be anywhere near the truth. Biela's Comet is dead; it broke in half in 1846 and has never been seen since 1852, and the meteors representing its debris are now so few that the shower is virtually undetectable. Phaethon is not a comet at all, but an asteroid. Its orbit is very like that of the Geminid meteor stream, and it is often supposed that Phaethon itself may be an old comet which has lost all its volatiles. I admit to being dubious, but it is at least a possibility.

But unless we include Lexell's Comet, which has been so roughly treated by Jupiter, all the parent comets listed here, apart from Thatcher's, have relatively short periods. We see Lyrids every year, even though the ZHR is variable. This means that there must be debris spread all round the orbit, and this in turn means that Thatcher's Comet has made many revolutions round the Sun. Since it takes over four centuries to complete one orbit, we may therefore assume that it is very ancient indeed.

I wonder what Thatcher's Comet will look like when we see it again, around 2276? Readers of this book will never know, but perhaps our descendants will be there to welcome the comet back.

38

Apocalypse Postponed

On 13 March 1998, the London *Mirror* published some remarkable headlines. The article, on page 7, was entitled "End of the World?" with a subtitle: "At 6.30 p.m. on October 26, 2028, a mile-wide asteroid with the destructive power of two million Hiroshima bombs is coming". It went on as follows:

"A mile-wide asteroid that could collide with Earth 30 years from now is threatening to wipe out civilisation, scientists revealed yesterday.

"The impact, at more than 17,000 mph (27,000 kph), would cause an explosion equivalent to two million Hiroshima size bombs.

"These would be the consequences of the space rock catastrophe 'timetabled' by astronomers for 6.30 p.m. on October 26, 2028.

"Huge earthquakes would be triggered and the blast heat would cause a volcano-like meltdown.

"Dust clouds would blot out the Sun, plunging the world into a prolonged cosmic winter.

"Millions would die in a total collapse of society.

"British scientist Dr Benny Peiser said of the asteroid code-named 1997 XF11 yesterday: 'It would not necessarily mean the end of mankind, but it would wipe out civilisation as we know it. We would regress to the level of the Dark Ages. All the trappings of modern life would be totally gone.'"

There can be no doubt at all that the impact of an asteroid of this size would be devastating and, not unnaturally, there was a great deal of popular interest. (I lost count of the number of broadcasts I made over the next forty-eight hours.) From the start I was decidedly sceptical because the data were blatantly inadequate to lead to any definite conclusions.

Jim Scotti, of the University of Tucson, working with the space telescope at Tucson, Arizona, discovered the asteroid on 6 September 1997. It was moving in a way that showed that it must be relatively close, and Brian Marsden and Gareth Williams, of the Harvard-Smitheibian Center for Astrophysics at Cambridge, Massachusetts, added it to their list of "potentially hazardous objects", which could conceivably hit the Earth. From the outset there was considerable uncertainty, but in an electronic circular sent out on 11 March Brian Marsden announced that there would probably be a very close approach.

What has to be remembered is that for a reliable orbit to be calculated, an asteroid, or any other object, must be observed for a reasonable period of time. Yet there were very few observations of 1997 XF11, and what was needed was an earlier sighting dating back for several years. This was the reason for Marsden's circular. He was anxious to obtain more information, and it seemed that the asteroid would have passed by us in 1957, 1971, 1976, 1983 and 1990. An examination of plates taken at those times might show up an image.

So it proved. Eleanor Helin, probably the world's most skilled asteroid-hunter, tracked it down on a photograph exposed in 1990. At once it became possible to revise the orbit, and it became clear that, instead of hitting us, 1997FX11 would pass by at a distance of at least 600,000 miles (970,000 km).

It was Marsden's circular that sparked off the furore, but no possible blame can be attached to Marsden himself; all he wanted to do was obtain further observations — and he did. All the same, there was a lesson to be learned, and it has now been tacitly agreed that no general information will be released about these space bullets until a proper orbit is known.

39

The Star in the East

Every year, as Christmas approaches, one question is asked time and time again. "What was the Star of Bethlehem?" Assuming that it has some scientific explanation and is not merely a charming story (which is by no means certain), can we pin it down?

I have made numerous broadcasts about it, but I have to make one point clear from the outset. I cannot tell you what the Star of Bethlehem was. What I can do, at least I believe so, is to tell you what it wasn't. But first let us go back to our only source of information — the Bible — and see what we have to guide us.

The plain truth is "very little". The Star is mentioned only once, in the Gospel according to St Matthew, chapter 2. Matthew writes that the wise men from the East came to Herod, saying "Where is he that was born King of the Jews? for we have seen his star in the east, and are come to worship him." Verses 7 to 10 run as follows:

"Then Herod, when he had privily called the wise men, enquired of them diligently what time the star appeared. And he sent them to Bethlehem, and said: 'Go and search diligently for the young child; and when ye have found *him*, bring me word again, that I may come and worship him also.'

"When they had heard the king, they departed; and lo, the star, which they saw in the east, went before them, till it came and stood over where the young child was. When they saw the star, they rejoiced with exceeding great joy."

St Matthew says no more; the other Gospels do not mention the Star at all, so that from the very beginning our information is depressingly meagre. To make matters worse, we are by no means sure about our dates. The one thing we do know for certain is that Christ was not born on 25 December, AD 1. Our AD dates are reckoned according to the calculations of a Roman monk, Dionysius Exiguus, who died in the year we now call AD 556.

He worked out that the birth of Christ occurred 754 years later than the founding of Rome, and the system has become so firmly established that it will never be altered now, even though it is definitely wrong; Christ must have been born several years earlier than AD 1. Note that there was no year 0, which is why the start of the forthcoming Millennium will be on the first day of 2001, not the first day of 2000. Quite recently the Italian scholar Giovanni Baratta has pointed out that Dionysius failed to take this into account, and also omitted the four year period when the Roman emperor Augustus ruled the Imperial City under his original name of Octavian, as well as the first two years of Augustus' successor, Tiberius. All this brings the birth of Christ back to 12 BC, which most scholars believe to be too early. Moreover, 25 December was not celebrated as Christmas Day until the fourth century, by which time the real date had been forgotten, so that our Christmas is wrong too. Clearly, Biblical historians are batting on a very difficult wicket.

If we want to find a scientific explanation for the Star, we must look for something in the sky, which was:

1. Very brilliant — so brilliant that it could not have been overlooked.
2. Seen by only the Wise Men, and nobody else, so that it must have been
3. Very short-lived. Also:
4. Very unusual, as by the time of Christ the sky was well known to all "educated" people.
5. Quick-moving.

So far as superficial appearance is concerned, the planet Venus would fit our requirements quite well. It can be truly magnificent, looking like a small lamp in the sky and capable of casting shadows. Moreover, it is at its best when seen in the east before sunrise (or in the west after sunset). It must therefore have been a brilliant morning star reasonably often during the period that covers the appearance of the Star of Bethlehem. Unfortunately, we can show straight away that the Star was not Venus or any other planet. Two thousand years ago, the apparent motions of the celestial bodies were well known, and if the

Star had been Venus, everybody would have known about it —
indeed, Herod himself would have had to do no more than go
and look. If the Wise Men were deceived by Venus, they would
not have been very wise. The same argument disposes of Jupiter, Mars and all the other planets as well as the brightest stars,
such as Sirius and Canopus.

The usual "red herring" is what is termed a planetary conjunction. There are times when two planets seem to lie close
together in the sky. Of course, this is merely a line of sight effect, with one planet in the foreground and the other much further away, but it can be spectacular. It has even been known for
one planet to pass in front of another and hide or occult it. For
example, on 3 January 1818 Venus occulted Jupiter, and this
will happen again on 22 November 2065, though both planets
will be rather inconveniently close to the Sun in the sky. There
was a conjunction of Jupiter and Saturn in the year 7 BC, and
this is always being trotted out as an explanation of the Star.
However, this whole idea is definitely out of court because the
conjunction would not have been really dramatic; it would have
lasted for some time, and it would have been widely observed,
so that it fails to satisfy any of the criteria in our list apart from
the fourth.

The next idea involves a nova or a supernova. A nova occurs when a formerly dim star suddenly flares up to many times
its normal brilliancy, but again the outburst lasts for some time
and no nova was reported by contemporary astronomers. Only
one possibility seems to remain: a meteor. Meteors are tiny particles, usually smaller than pins' heads, moving round the Sun.
They are cometary debris, and are common enough — nearly
everyone must have seen a shooting star at one time or another.
An eastward-moving meteor might well have flashed across the
sky above the Wise Men, and it might have been very bright —
now and then we see meteors which are more brilliant than the
Moon. It is also possible that a second meteor may have been
seen later, travelling in the same direction. A meteor (or meteors) would satisfy all the criteria I have listed. Of course it does

not explain how the Star could hover over the place where the baby Christ lay. If the meteor explanation is valid, we have to make due allowance for poetic licence.

I would be the last to claim that my idea is fully convincing but I do suggest that it is rather better than the rest. We will never know the truth now.

40

It Was in the Papers!

Generally speaking, the daily press is not noted for its scientific accuracy. Of course, there are some good science reporters, but sometimes one comes across very curious statements. There are also very correct reports of very odd occurrences. The following is a more or less random selection, very roughly chronological. I have avoided most flying saucer accounts (the Roswell hoax alone would fill many pages) and have said relatively little about astrology, except where it has impinged upon everyday life. I have also excluded papers such as the *Sunday Sport*. I cannot claim that I have personally checked all these reports, but at least I hope that some of them will entertain you.

"A large meteor is due to crash into Southern Britain in the next few years." This claim is made by Mr L. Ferret, Chairman of the ISWG (Island Sky Watchers Group). In an exclusive interview Mr Ferret explained when and how the disaster will happen. "I have been tracking a large meteor heading toward earth for some time, and believe it will strike the Island between 8.30 a.m. and 10.00 a.m. on 14 February 1995. It is estimated to be 20 times bigger than Haley's* Comet. ... Ground Zero will be the Country Council Offices in Newport." The ISWG has a detailed plan on how to save the island from destruction. These plans will be published in our next issue. So watch this space; your life could depend on it. *Wight Pages*, 6 October 1993.

* His spelling. · P.M.

Hailed as Mankind's most important hour, it drew a shivering crowd of only 300. They huddled outside the cathedral of St Sophia (Ukraine) to witness the end of the world, as predicted by the prophet Marina Tsvygun. But after several hours of not very much happening in a temperature of -23° F, they all went

home. It was left to Ukrainian Interior Ministry spokesman Valentyn Nedrehaylo to declare the obvious: "Today is off. The end of the world is hereby cancelled."
Daily Mail, 15 November 1993.

John Whitmore, BSc (University of Kent State), MA, is Assistant Professor of Geology at Cedarville College, Ohio. ... He is not a crank (although some think he is) but a creationist scientist.

His views are as follows: God created the heavens and the earth about 6000 years ago. Creationists accept that the Book of Genesis is a literal description of the beginning of the world, starting with night and day on day one and ending with God taking a well-earned rest on the following Sunday.

Dinosaurs arrived between day four ("let the waters abound with an abundance of living creatures") and day five ("let the earth bring forth the living creature according to its kind"). Man and woman followed later that afternoon.

For the next 2000 years the descendants of Adam and Eve lived with the dinosaurs and all other living creatures. Most perished in the Flood, but two of every species were saved in Noah's Ark. ... The earth we know today was shaped by the Flood, earthquakes, volcanoes, and "fountains of the great deep". The fossils we find today were buried in sediment.

After the Flood, the animals from the Ark multiplied, but were now allowed to be hunted by Man and by each other. Dinosaurs became extinct some time during the past 2000-3000 years, probably by the hand of man. Small numbers may survive (e.g. the Loch Ness Monster). *Sunday Telegraph*, 16 February 1997.

General Chatichai Choonhavan's astrologer had a decisive influence on the formation of Thailand's new government, which was sworn in yesterday by King Bhumibal.

The astrologer, whose identity is secret, told the Prime Minister that it was vital for his new Government to be announced last Friday. He complied, but the hasty procedure led to the exclusion of a leading party from the new coalition.
Times, 17 December 1991.

Astronauts on the Space Shuttle *Endeavour* peered yesterday at tiny tadpoles hatched in space. Their weightless aquatics were "certainly not what one would see on earth," said Ken Souza, a NASA scientist. "They were swimming in backward somersaults, forward somersaults. Some froze, some swam naturally."
Independent, 12 June 1993.

Leaders of a South Korean Christian sect apologised yesterday for failing to arrange assumption to heaven for thousands of worshippers who had prepared to depart Earth in a comet-like flash called "the rapture".

Most leaders of the Mission for the Coming Days Church, which predicted Doomsday for midnight on Wednesday, said the timing had been wrong. They said they would be in touch with Jesus later.
Daily Express, 27 October 1993.

"I must query the statement in the June *Sky at Night* programme that the Earth spins round . . . the Earth is located at the centre of the universe, and it is the universe doing the moving.

"Copernican theory led to the English, French and American revolutions, Darwin, Lyell, Huxley, Marx, Nietzsche, atheistic existentialism, materialistic hedonism, despair and even Nazism and modern racism."
Letter from Amnon Goldber in *The European*, 17 June 1993.

A variety of artefacts from the Soviet space programme, including some that are still in space, sold for a total of nearly $7 million (£4.7 million) in New York.

One bidder at Sotheby's on Saturday paid $68,500 for the Lunokhod 1 surface rover and the Luna 17 descent stage, both of which are still on the Moon's surface. The buyer has not said when or how he intends to collect his purchase.
Times, 18 December 1993.

Customers are flocking to buy American scientist Barry Skinner's untested invention — a £100 alien detector which shrieks

if a creature from outer space comes near.
News of the World, 14 February 1994.

Torstein Sele has converted an old NATO rocket launcher into a muck-spreader for his farm in Norway. *Sun*, 16 April 1994.

British Airways is trying to trace a list of reservations made by 3000 children and adults on the inaugural lunar shuttle. The lost "Moon list" was drawn up between 1968 and 1974 by BA's predecessor, the British Overseas Airways Corporation. Reservations can be handed down through wills. *Times*, 3 July 1994.

When he hit that golf shot on the Moon 23 years ago, a fortune was eschewed by Alan Shepard (America's first space man in 1961) because he refused to divulge the name of the ball — and he's still not telling! *Daily Telegraph*, 5 July 1994.

Another space "first" takes place in February, when the space shuttle carries chilled Coca-Cola and Diet Coke into orbit. Astronauts will be asked to give opinions on the flavour and carbonation. *New Scientist*, December 1994.

Farmer Howard Davies and his wife Margaret are sitting on an egg that could be worth £10,000. That is the prize in a contest to find the unusually marked egg which, according to folklore, is laid whenever Halley's Comet passes Earth. Mrs Davies, 42, of Pontardawe, Swansea, said yesterday: "Ours has a near-perfect circle with a halo round it — just like a planet surrounded by haze." Sadly, date lost.

"Sir, I never cease to marvel at the sense of purpose and design evident in the universe. How is it, for example, that the fragments of a distant comet, invisible to the naked eye, could crash into Jupiter in strict alphabetical order?"
Letter in the *Times*, 17 July 1994.

A firm in Chita, Siberia, is inviting customers to insure themselves for 100,000 roubles against being kidnapped by aliens.
They advise clients taken on space voyages by extra-terrestrials
to have their travel certificates stamped by their hosts, because
those who make claims may have to prove them in court.
Daily Express, 22 October 1994.

Frightened viewers jammed a TV network's switchboards after
a fake broadcast showing huge asteroids crashing into Earth.

The Halloween show *Without Warning* was broadcast in the
US 56 years after Orson Welles' famous radio dramatisation of
The War of the Worlds by H.G. Wells. His vivid portrayal, on 30
October 1938, of a Martian invasion of Earth fooled hundreds
of Americans and almost sparked mass panic.

Sunday's programme broadcast "live" reports from disaster sites
around the world. Viewers were shown scenes of panic and death
from cities in Europe, Asia and the US, and told the world's
stock markets had collapsed. American F-16 fighters were shown
being scrambled to destroy the incoming asteroids with nuclear
missiles. Later a CBS news presenter apologised; saying that
Without Warning may have confused some viewers.
Daily Mail, 1 November 1994.

Switzerland has rejected an application to build an embassy for
visitors from outer space. AFP.
Daily Telegraph, 15 November 1994.

The Roman Catholic Church, which uses Latin for many official
documents and writings, is to publish the first truly modern dic-
tionary in decades. The new tome contains some notable surprises — flying saucers (*coruscantes disci per convexa coeli volantes, or
clipei ardentes*) make a rare appearance in the language of Cicero
and Caesar. *Daily Telegraph*, 5 January 1995.

British inventor James Langford has just offered NASA an oven
for the space shuttle. It has a revolving drum and resembles a
washing machine. *Daily Mail*, 24 March 1995.

Astronauts aboard the space shuttle *Endeavour*'s trained telescopes on the constellation Sagittarius to help astronomers study binary stars. *Daily Telegraph*, 6 March 1995.

Tourists in Russia are paying £4600 a head for five days' training as cosmonauts. *Sun*, 25 April 1995.

If your sports car is too slow and your private jet flies too low, then Victor Rylkov has just the status vehicle for you: a fully operational second-hand Russian space shuttle in mint condition — a snip at $10 million.

An advertisement appeared in the classified section of the *New York Times* this week offering to sell the 70-ton Buran spacecraft to any "serious" buyer. Mr Rylkov can also offer another Buran, unsuitable for space flight, and hopes to have two more shuttles available soon. If he is satisfied with the buyer's intentions, he says he can deliver within four to six weeks. *Times*, 2 June 1995.

Suggestions from the bizarre to the brutal have flooded in to NASA from people eager to prevent a woodpecker from delaying the multi-billion dollar American space programme.

The ideas came after the 21st launch of the shuttle *Discovery* had to be postponed because a yellow-shafted flicker woodpecker had pecked more than 75 holes in the insulating foam around the fuel tank, apparently in search of a place to mate. The shuttle, which should have lifted off yesterday, had to be wheeled back into its hangar at a cost of £62,893.

Practical suggestions included painting the tanks blue, which is apparently loathed by woodpeckers, using jet engines to engineer a draught above the tanks, knocking out the birds with stun guns, and draping protective nets over the shuttle.

More innovative remedies included boiling cabbage and spraying the water on to the tanks, sprinkling vulnerable areas with raccoon scent, and using a shaman to summon helpful spirits. One writer felt that the problem could be handled with a well-aimed rifle, ruled out because the launch pad is in a nature reserve.

For now, the responsibility for protecting the shuttle has fallen to several strategically placed plastic owls and a high-pitched siren. *Times*, 9 June 1995.

International companies are being invited to bid upward of £625,000 to place an advertisement on the side of a £60 million European Space Agency rocket to be launched in November by the Swedish Space Corporation. The message will be visible with the naked eye for 10 seconds after take-off and the exploratory space-flight will last for 15 minutes. *Times*, 14 August 1995.

Mysterious disks of light in the night sky that prompted a UFO report in Torquay, Devon, have been traced to a garden fish-pond. The light was reflected from the ultra-violet filtration system of Joseph Martin's pond at his home in Woodland Park. *Daily Telegraph*, 7 September 1995.

It is the achievement of a dream for determined UFO watchers. But more sceptical earthlings are bug-eyed over a bizarre scheme from Brussels for spending even more money.

A report for the European Parliament will recommend that the EU establish an office to track UFO sightings. In short, Brussels wants to set up its own X-files investigation into flying saucers and little green men. ...The Defence Ministry has in the past limited its discussion of the case to assurances that British security was not threatened. Yesterday it declined to comment. *Daily Mail*, 5 September 1995.

The extent of the cash crisis in Russia's space programme can now be revealed — caviar and other specialities have been removed from the cosmonauts' menu. The cutback came after the space agency extended the current Mir space station mission by 5 weeks because it cannot afford a rocket to take up a new crew. *Daily Mail*, 20 October 1995.

Millions of Cambodians will be shooting at the sky during a total eclipse later this month because of an ancient tradition which

warns that bad luck will follow if the Moon "eats" the Sun. "They will try to help the Sun escape," said an observer.
Daily Express, 11 October 1995.

India's Prime Minister, P. V. Narashima Rao, has embarked on an extended foreign tour to avoid the malign influences of a total solar eclipse which will strike his house just after 8.30 a.m. tomorrow. His astrologers, whose advice he heeds to the letter, told him to leave the country several days before the eclipse and to remain abroad for a long time afterwards. *Times*, 23 October 1995.

Amazed scientists who thought they were picking up messages from outer space were really getting signals from a microwave. The US researchers reckoned they'd stumbled on the biggest space prize of all when their 64-metre radio telescope recorded a regular "beep". It came at the same frequency at the same time each evening and because it was recorded by high-tech wizardry programmed to erase anything earthbound, the team were convinced they were in touch with aliens.

In their excitement, they forgot one thing. The little green men always beamed in at dinnertime.

Far from originating from new life forms in the depths of space, the signals were coming from a microwave in the kitchen downstairs. And what the scientists thought was a message saying "Take me to your leader" was really a reminder to "Take me to your larder."

The experts from California's SETI Institute admitted their blunder at a top conference. Embarrassed Institute member Peter Backus told the American Astronomical Society in Texas, "The signals were all coming from our own technology."
Daily Mirror, 18 January 1996.

Montpellier: police and psychiatrists were helping a man who sought their aid in tracing his lost vehicle — a space ship on which he claimed to have landed 350 years ago. He said he needed to return to his planet, as he had left his guitar there.
Daily Telegraph, 27 January 1996.

A satellite that went missing a year ago has turned up again in the bush of northern Ghana. German and Japanese scientists who thought it had plunged to destruction in the Pacific are delighted. The £28 million Express satellite was launched on a Japanese rocket in January 1995. It went into the wrong orbit, lost contact and crashed, to the chagrin of scientists.

Cut to the northern Ghanian bush, where local people later found a re-entry capsule with parachute attached. The parachute was Russian made and contained Cyrillic lettering, so the authorities in Ghana feared it might be radioactive. They stuffed it in a cupboard at a nearby airport.

Then a German diplomat read an article about the mystery capsule and put two and two together. A team from the German space agency has confirmed that the satellite is theirs. Now all that remains is for them to get it back. *Times*, 4 February 1996.

William Hill bookmakers have laid odds of 14,000,000 to one that a UFO piloted by Elvis Presley will crash on to the Loch Ness Monster. *Sun*, 25 February 1996.

American scientists have worked out why a satellite they launched with Russia failed after only one day in orbit. The Russians had wired up the solar panels the wrong way round.
Daily Telegraph, 22 March 1996.

A Chinese farmer's discovery may mean that dinosaurs never disappeared from the Earth at all, but adapted to new conditions and became birds. The find, of a small-feathered dinosaur, has been dated as 120,000,000 years old.
Independent, 27 September 1996.

According to Professor Peter Usher of Pennsylvania State University, Shakespeare's *Hamlet* is nothing more nor less than a cosmological treatise. "Hamlet is an allegory for the competition between Thomas Digges of England and Tycho Brahe of Denmark"... Shakespeare knew Digges, Professor Usher says, and through him also knew the Danish astronomer Tycho Brahe, whose cosmology was Earth-centred. "When Hamlet states 'I could be bounded in a nutshell and count myself a king of infinite space,' he is contrasting the shell of fixed stars in the Ptolemaic and Tychonic models with the Infinite Universe of Digges. Claudius is named for Claudius Ptolemy, who perfected the geocentric model, while Rosencrantz and Guildestern personify Tychonic geocentricism. Thus when Rosencrantz and Guildenstern are killed, so are Tycho's ideas, and when Claudius is killed it signals the end of geocentricism. The chief climax of the play is the return of Fortinbras from Poland and his salute to the ambassadors of England. Here Shakespeare signifies the triumph of the Copernican model and its Diggesian corollary. Copernicus was a Pole." *Times*, 14 January 1997.

Aliens are among us and are under Satanic control, according to UFO Concern, a predominantly Anglican pressure group founded by Lord Hill-Norton, a retired Admiral of the Fleet and former Chief of Defence Staff. *Daily Telegraph*, 28 February 1997.

Astronauts will make a fifth spacewalk from the shuttle *Discovery* tonight to patch up the Hubble telescope's silvery foil protective skin with sticky tape. *Daily Mail*, 17 February 1997.

A Capital Radio DJ announced that because of the adjustment between GMT and BST, April 5 and April 12 had been cancelled. *Sun*, 20 March 1997.

France was thwarted yesterday in an attempt at a meeting of European Union transport ministers to scrap summer time. It was outvoted by other EU nations which argued that studies had found that changing the clocks brought energy and other savings of about £12 billion a year.
Daily Telegraph, 12 March 1997.

The General Election coincides with a powerful omen of change in the form of the Hale-Bopp comet, says *Daily Mail* astrologer

Jonathan Calner. "Without it I'd predict a Labour landslide. With it, Tony Blair could win by a narrow majority and need Paddy Ashdown's help to form a Government."
Daily Mail, 19 March 1997.

A brewer has named his latest beer after the Hale-Bopp comet. Ale Bopp, brewed by Alan Thomson, of the Old Chimneys brewery in Market Weston, Suffolk, will be in selected pubs until the comet leaves the Solar System. The beer, said to be very hoppy, will be launched at the Bury St Edmunds beer festival today.
Times, 11 April 1997.

One of Britain's foremost scientists is suing the makers of a leading brand of lavatory paper. Sir Roger Penrose, Rouse Ball Professor of Mathematics at Oxford University, says that although Kleenex Quilted may be super-absorbent and gentle on the skin, it has no right to achieve this by using the mathematical pattern he designed 20 years ago. In his legal action he seeks destruction of all stocks of the paper bearing this pattern.
Times, 12 April 1997.

The odds against meeting intelligent extraterrestrials by January 1, 2000 were cut by bookmakers William Hill yesterday to 33-1. A spokesman said: "We've been lowering the odds for a few months. They were 1000-1 at one point."
Daily Mail, 14 June 1997.

Headline in the *Daily Telegraph*, 24 July 1997:
LIQUIDATOR TO TARGET GALILEO'S ADVISERS.

A lawsuit has been filed against NASA, the American space agency, to prevent it revealing any more details about conditions on Mars.
   Three men, from Yemen, claim that they own Mars and that their permission should have been sought by NASA before its *Pathfinder* spacecraft landed on July 4 and released a buggy called

Sojourner which has been testing the planet's rocks and soil, sending back photographs which reveal a landscape as inhospitable as the Arizona desert.

But the Yemeni men, Adam Ismail, Mustafa Khalil and Abdullah al-Umari, claim the unlikely rockscape was home to their ancestors thousands of years ago, according to the weekly *Al Thawri* newspaper. They want NASA to suspend releasing the results of any further explorations, and any information about the Martian atmosphere, surface or gravity, until the court case is over. The men said in documents presented to Yemen's prosecutor-general: "We inherited the planet from our ancestors who lived on it 3000 years ago."

NASA showed no sign of suspending its activities yesterday, announcing that soil on Mars is made up of a silt that is finer than talcum powder. *Daily Telegraph*, 25 July 1997.

Well, these are some of the reports which I have collected during the past few years. Inevitably most of them come from the newspapers which I happen to take daily, but there must be many others. If you come across any, do please let me know about them for a new edition of *The Wandering Astronomer*!

41

Caribbean Eclipse

There is no doubt in my mind that a total eclipse of the Sun is the grandest sight in all Nature. As the last sliver of the brilliant solar disc disappears, the sky darkens and the corona flashes into view. The spectacle is breathtaking. The trouble is that as seen from any particular location on Earth, total eclipses do not happen very often, because the Moon's shadow is only just long enough to reach terra firma. The track of totality can never be more than 169 miles (272 km) wide, so that you have to be in exactly the right place at exactly the right time. From England, there were no total eclipses between 1927 and 1999. The next will be delayed until September 2090.

According to Spode's Law — "if things *can* be awkward, they *are*" — totality tracks usually cross areas which are either well away from land or are extremely difficult to get at. And, of course, there is always the danger that the sky will be cloudy at the critical moment. I am fortunate enough to have seen several totalities, and three of these have been from the sea. Eclipse-chasing by ship has the obvious advantage that if the weather prospects are poor, you can — within certain limits — move to a clearer area. This happened in 1995 when I was on board the cruise ship *Marco Polo*. Our Norwegian captain manoevred the ship into the only cloud-free spot for many miles around.

For the eclipse of 26 February 1998, I was again scheduled to be afloat; this time on the *Stella Solaris* which flies the Greek flag and has a Greek captain. My rôle was twofold. I am not a proper observer of the Sun — the Moon is my main interest — but I was due to give a few lectures to the ship's passengers, and also to make a television broadcast for the BBC *Sky at Night* programme during the eclipse itself.

One particularly enjoyable feature of an eclipse cruise is that there are so many congenial people around! On the *Stella Solaris* we had as lecturers, Dr Edward Brooks of Boston College, the world's leading expert on eclipse weather; George Keene, whose knowledge of eclipse photography is second to none; Dr Paul Knappenberger, President of the Adler Planetarium in the US; Dr Edwin Krupp, Director of the Griffith Observatory in Los Angeles, who has pioneered the science of archeoastronomy; Dr Warren Young, also a Planetarium director; and Dr Ronald Parise, the only professional astronomer who was officially trained as an astronaut and who flew in 1990 aboard the Space Shuttle *Columbia* and in 1995 aboard *Endeavour*. We also had the *Sky at Night* team: Pieter Morpurgo to produce, Mike Winser to take the pictures and Doug Whittaker to look after the sound.

One very important person on the cruise was Dr Anthony Aveni, both in his capacity as an eminent astronomer and as the official "timekeeper" during the eclipse. This was a rôle I filled myself, I remember, during the eclipse of 1954 in Sweden. But Tony Aveni was far more efficient than I had been, so long ago. Finally, do not forget Ted Pedas, who was the project co-ordinator and who had organised the entire expedition.

What did we expect to see? Well, the usual phenomena of totality: the Diamond Ring, the chromosphere, the corona, the prominences and so on. The Sun was not long past the minimum of its eleven-year cycle of activity so we expected the corona to be more "spiky" and less symmetrical than when sunspots are plentiful. Several planets would, we hoped, be on view: Venus certainly, with Jupiter, Mars, Mercury and possibly Saturn. For this everything depended upon how dark the sky would become and this is something which can never be reliably predicted.

The *Stella Solaris* carried several hundred passengers, most of whom were keenly interested in the eclipse, though I must add that one dear lady solemnly read a book throughout totality! There was a full series of lectures: of special importance were those of George Keene, the photographic expert, because almost everyone was keen to take pictures — and the Sun is dangerous. Only during full eclipse is it safe to look at it directly through binoculars or a telescope. If even a tiny portion of the bright surface remains on view, the greatest care is needed. Either project the image, or else make sure that you are using the right types of filters. People tend to forget that a camera is a lens, and unprotected viewing through it will result in eye damage at best and permanent blindness at worst. Luckily, everyone took George's warning to heart and, so far as I know, there were no cases of "spots in front of the eyes".

I knew that I would be unable to take any pictures because I could not take part in photography and do a television commentary at the same time. Luckily Chris Doherty, my travelling companion, could be relied upon to take excellent pictures. I handed my camera over to a fellow passenger who aimed and clicked during totality and did, in fact, produce some highly creditable results.

Ed Brookes assured us that the weather would be kind and, as usual, he was right. With typical Greek seamanship, Captain Panarios took us to precisely the right place. There were no clouds to be seen and everything seemed to be going according to plan. Rehearsals took place on 24 February, and on the morning of eclipse day equipment was set up all over the decks. The BBC team chose an excellent vantage point on what was known as the Solaris deck, with Chris Doherty a few yards away. Also within range was Dudley Fuller, of telescope fame, whose main aim was to record the eclipse on video.

We were some way off the coast of Venezuela, well away from land. In fact the track crossed only a narrow neck of Central America — it had started in the Pacific and ended in the Atlantic so that the only land crossed, apart from Central America, consisted of a few Caribbean islands. (One of these was Montserrat, ravaged by the volcano and, to be candid, I was glad that we were not too near it!) Totality was due in the early afternoon and would last for only four minutes, so that everything had to be thoroughly rehearsed. Strange things can happen in the excitement of the moment. I remember that at one eclipse, from a ship off the coast of Africa, an enthusiastic astronomer took 40 photographs of the solar corona — with his lens hood in position the whole time . . .

The first contact came — a tiny notch in the edge of the Sun.

Slowly it spread until the light dimmed and the temperature began to fall. Final preparations were made. Mercifully, the *Stella Solaris* was rock-steady; Captain Panarios had seen to that. Then, suddenly totality was upon us. The Moon's shadow raced over the sea; there was the flash of the Diamond Ring, together with the co-called Baily's Beads, due to sunlight coming toward us through valleys on the Moon's rough edge; and then came the corona.

It was without doubt the loveliest total eclipse in my experience. The corona was brilliant and structured, and there were two bright red prominences. Nature seemed to go into a state of suspended animation. Tony Aveni broadcast the time — otherwise I suppose the only other voice was mine, carrying out my commentary loudly enough to satisfy Doug Whittaker but, I hope, not loudly enough to disturb the viewers. Those four minutes seemed more like four seconds. Then, again, the Diamond Ring, more glorious than ever. The corona faded, the light flooded back over the sea and, in an instant, Nature seemed to wake up as suddenly as she had gone to sleep. Instinctively there was a tremendous burst of applause. The Sun — and Captain Panarios — had served us well.

Then the aftermath. With the BBC team I went round to see how various people had managed. Chris Doherty was highly delighted: his plan of campaign had worked perfectly. So was Dudley Fuller, whose video had, he thought, been successful (as indeed it was). George Keene had taken his usual splendid photographs, and all the various investigators had been very happy indeed. Someone even saw those elusive features, the shadow bands; and of course the planets, plus a few bright stars, had shone forth.

Astronomically, that was that. The cruise itself went on for another nine days, taking us to Curacao, Aruba, Jamaica, the Cayman Islands and Cozumel in Mexico, before we disembarked at Galveston in Texas to fly home. It was all most pleasant. We even put on what we hoped was entertainment. Dudley Fuller is a brilliant jazz pianist, while I hope that I did not disgrace myself playing the xylophone. Obviously we had to wait until we could have our photographs developed, but in the event they did come up to expectations.

Yes, I thoroughly recommend eclipse cruising. And in my own case, there was the added bonus that on 4 March, a day before we had to leave the *Stella Solaris,* I had my seventy-fifth birthday. To my genuine surprise, I was the guest of honour at a party on deck — even with a cake! — and although I do feel decidedly antique, what could be better than passing through this milestone in one's life than to celebrate it on a cruise ship in the Caribbean?

Glossary

Absolute Magnitude The apparent magnitude which a star would have if it were observed from a standard distance of 10 parsecs (32.6 light years).

Albedo The ratio of the intensity of light reflected from an object, such as a planet, to that of the light it receives from the Sun.

Ångström Unit One hundred-millionth part of a centimetre.

Apparent Magnitude The apparent brightness of a celestial body; the lower the magnitude, the brighter the object. The magnitude of the Sun is about -27; the Pole Star +2, normal limit of naked-eye visibility +6; faintest detectable with modern equipment around +29.

Astronomical Unit The mean Earth-Sun distance: 149,598,000 km (approximately 93,000,000 miles).

Binary Star A star made up of two components, moving round their common centre of gravity.

Black Hole A region round an old, collapsed star from which not even light can escape.

Celestial Sphere An imaginary sphere surrounding the Earth, whose centre is coincident with the centre of the Earth.

Cepheid A short period variable star; the variations are very regular. Their real luminosities are linked with their periods; the longer the period, the more luminous the star. The name comes from the prototype star, Delta Cephei.

Corona The outer part of the Sun's atmosphere.

Cosmology The study of the universe considered as a whole.

Culmination The maximum altitude of a celestial body above the horizon.

Day, Sidereal The mean interval between successive culminations of the same star: 23h 56m 4s.091.

Declination The angular distance of a celestial body north or south of the celestial equator. It corresponds to latitude on Earth.

Direct Motion Movement of revolution or rotation in the same sense as that of the Earth.

Doppler Effect The apparent change in wavelength of the light from a luminous body which is in motion relative to the observer. With an approaching object the wavelength is effectively shortened, the object appears "too blue", with a receding object the wavelength is lengthened, and the object appears "too red". This affects the positions of the lines in a stellar spectrum; the amount of blue or red shift gives a key to the star's velocity.

Double Star A star made up of two (or more) components — either genuinely associated (binary pair) or merely lined up (optical pair).

Ecliptic The projection of the Earth's orbit onto the celestial sphere. It may also be defined as the apparent yearly path of the Sun against the stars.

Equator, Celestial The projection of the Earth's equator on to the celestial sphere.

Equinoxes The two points at which the ecliptic crosses the celestial equator; crossing of the Sun from south to north (Vernal Equinox or First Point of Aries, around March 21 each year) and from north to south (First Point of Libra, around September 22 each year). Owing to precession, the vernal equinox is now in Pisces and the autumnal equinox in Virgo.

Escape Velocity The minimum velocity that a body must have if it is to escape from a planet, or other body, without being given extra impetus. The Earth's escape velocity is 7 miles (11 km) per second.

Extinction The apparent dimming of a celestial body when low over the horizon, so that its light comes to us via a denser layer of the Earth's atmosphere.

Faculae Bright, temporary patches above the Sun's photosphere.

Flares Brilliant eruptions in the outer part of the Sun or any other star.

Fraunhofer Lines The dark absorption lines in the spectrum of the Sun.

Galaxies Systems made up of stars, nebulae and interstellar matter.

Galaxy, The The Galaxy which contains our Sun. It includes about 100,000 million stars.

Gamma Rays Radiations of extremely short wavelength.

Globules Dark patches inside nebulae; they are probably embryo stars.

H.I and H.II regions Clouds of hydrogen in the Galaxy; H.I if the atoms are neutral, H.II if they are ionised.

Halo, Galactic The spherical-shaped cloud of stars round the main Galaxy. Globular clusters are halo objects.

Heliacal Rising The rising of a star or planet at the same time as the Sun, though the term is generally used to denote the time when the object is first visible in the dawn sky.

Hertzsprung-Russell (H-R) diagram A diagram in which the stars are plotted according to their spectral type and absolute magnitude.

Hubble Constant The rate of increase in the recessional velocity of a galaxy with increased distance from Earth. Its current value is uncertain.

Interferometer, Stellar An instrument for measuring star diameters. It is based on the phenomenon of the interference of light-waves.

Ion An atom which has lost one or more of its planetary electrons, so that it has a positive charge of electricity.

Kiloparsec One thousand parsecs (3260 light-years).

Libration The apparent "tilting" of the Moon as seen from the Earth, due to the Moon's changing orbital speed.

Limb The edge of the apparent disk of the Sun, a Moon, or a planet.

Light-year The distance travelled by light in one year: it is equal to 9.4607 million million kilometers.

Local group A group of more than two dozen local galaxies. It includes the Milky Way system, the Magellanic Clouds and the Andromeda Spiral.

Lunation The interval between successive new moons: 29d 12h 44m.

Main Sequence A band on the H-R Diagram from top left to bottom right.

Maria Plural of mare, plains on the surface of the Moon.

Megaparsec One million parsecs.

Meteor Cometary debris, a tiny particle which burns away in the Earth's upper air.

Meteorite A body which comes from the asteroid belt and lands on Earth. Meteorites may be irons, stones or a mixture of both.

Nebula A cloud of gas and dust in space.

Neutrino A fundamental particle with little or no mass and no electrical charge.

Neutron star The remnant of a very massive star which has exploded as a supernova. Neutron stars are small and incredibly dense.

Newtonian reflector A telescope in which the light is collected by a curved mirror and sent to the eyepiece via a small flat mirror in the upper part of the tube.

Nova A star which suddenly flares up; the outbreak takes place in the highly-evolved component of a binary system.

Occultation The covering-up of one celestial body by another.

Opposition The position of a planet or other body in the sky when it is exactly opposite to the Sun.

Orbit The path of a celestial body.

Parsec The distance from which a body would show an annual parallax of one second of arc: 3.26 light-years, 206,265 astronomical units or 30.857 million million kilometres.

Penumbra (1) The arc of partial shadow to either side of the main cone of shadow cast by the Earth. (2) The lighter part of a sunspot.

Perigee The position of the Moon or other body in orbit round the Earth, when closest to the Earth.

Perihelion The position of a body orbiting the Sun when at its closest to the Sun.

Period The time required by a body to make one complete rotation on its axis.

Perturbations The disturbances in the movements of a celestial body caused by the gravitational pulls of other bodies.

Phases The apparent changes in the shape of the Moon and planets due to the differing amount of sunlit side turned Earthward.

Photosphere The bright surface of the Sun.

Planetary nebula A small, hot, dense star surrounded by a shell of gas. It is not truly a nebula and has nothing to do with a planet!

Poles, Celestial The north and south points of the celestial sphere.

Position angle The apparent direction of one object with reference to another, measured from north through east, south and west.

Prominences Masses of glowing hydrogen rising from the Sun's surface. They were formerly, and quite incorrectly, termed Red Flames.

Proper motion, Stellar The individual motion of a star on the celestial sphere.

Quadrature The position of the Moon or a planet when at right-angles to the Sun as seen from Earth.

Quasar A very remote, super-luminous object, now thought to be the core of a very active galaxy. (Also known as a QSO or Tuasi-Stellar Object.)

Radial velocity The toward-or-away movement of a star relative to the Earth.

Radiant The point in the sky from which the meteors of any particular shower appear to come.

Retrograde motion Orbital or rotational motion in the sense opposite to that of the Earth. (I have likened it to the movement of a car going the wrong way round a roundabout!)

Right ascension The angular distance of a celestial body from the vernal equinox.

Scintillation Star twinkling. It is due entirely to the effects of the Earth's atmosphere.

Sidereal period The revolution period of a body round the Sun or a planet.

Solar wind A flow of atomic particles streaming outward from the Sun.

Solstices The times when the Sun is at its maximum distance (23½ degrees) from the celestial equator; around 22 June in the northern hemisphere and 22 December in the southern hemisphere.

Superior planets All planets lying further from the Sun than we are.

Supernova A colossal stellar outburst. (Type I) the total destruction of the white dwarf component of a binary system. (Type II) the result of the explosion of a star of high mass. Type II supernova produce neutron stars, which when detectable at radio wavelengths are known as pulsars.

Tektites Small, glassy objects found in localised areas, not now generally believed to come from the sky.

Terminator The boundary between the day and night hemispheres of the Moon or a planet.

Transit (1) The passage of a celestial body across the observer's meridian. (2) The projection of Mercury or Venus against the disk of the Sun.

Twilight The state of illumination after sunset when the Sun is less than 18 degrees below the horizon.

Vernal equinox The position where the Sun crosses the celestial equator, moving from South to north.

White dwarf A small, very dense star which has exhausted its nuclear "fuel".

Zenith The observer's overhead point.

Zenith hourly rate (ZHR) The number of meteors which would be expected to be seen by an observer under ideal conditions, with the radiant of the shower at the zenith.

Zodiac A belt stretching round the sky, 8 degrees to either side of the ecliptic, in which the Sun, Moon and planets (except Pluto) always lie.

Zodiacal light A cone of light rising from the horizon and extending along the ecliptic. It is due to thinly-spread interplanetary material being lit up by the Sun.

Index

Numbers in italics denote illustrations.

Picture Credits

Paul Doherty, 44, 94.
Hadfield, Cdr H.R., 2.
P.A. Holm, 21.
Hubble Space Telescope, 57.
Kronheim & Co. London, 9.
NASA, 45, 60, 61, 69, 156.
G. & W.B. Whittakers, T. Cadell & N. Hailes, London, 23.

Other photographs, artwork and diagrams have been sourced from the author's archives.